Studies in Computational Intelligence

Volume 1006

The series "Studies in Computational Intelligence" (SCI) publishes new developments and advances in the various areas of computational intelligence—quickly and with a high quality. The intent is to cover the theory, applications, and design methods of computational intelligence, as embedded in the fields of engineering, computer science, physics and life sciences, as well as the methodologies behind them. The series contains monographs, lecture notes and edited volumes in computational intelligence spanning the areas of neural networks, connectionist systems, genetic algorithms, evolutionary computation, artificial intelligence, cellular automata, self-organizing systems, soft computing, fuzzy systems, and hybrid intelligent systems. Of particular value to both the contributors and the readership are the short publication timeframe and the world-wide distribution, which enable both wide and rapid dissemination of research output.

Indexed by SCOPUS, DBLP, WTI Frankfurt eG, zbMATH, SCImago.

All books published in the series are submitted for consideration in Web of Science.

More information about this series at https://link.springer.com/bookseries/7092

Mohamed Alloghani • Christopher Thron
Saad Subair
Editors

Artificial Intelligence for Data Science in Theory and Practice

Springer

Editors
Mohamed Alloghani
Computing and Mathematical Sciences
Liverpool John Moores University
Liverpool, UK

The UAE Artificial Intelligence Office
Prime Minister's Office at the Ministry
of Cabinet Affairs and the Future
Dubai, UAE

Christopher Thron
Department of Science and Mathematics
Texas A&M University-Central Texas
Killeen, TX, USA

Saad Subair
College of Computer Studies
International University of Africa
Khartoum, Sudan

ISSN 1860-949X ISSN 1860-9503 (electronic)
Studies in Computational Intelligence
ISBN 978-3-030-92244-3 ISBN 978-3-030-92245-0 (eBook)
https://doi.org/10.1007/978-3-030-92245-0

This Springer imprint is published by the registered company Springer Nature Switzerland AG
The registered company address is: Gewerbestrasse 11, 6330 Cham, Switzerland

Preface

Before introducing the contents and purpose of this book, we first give some clarifying definitions.

Statistics has been an essential part of the pure and applied sciences for hundreds of years. Until the advent of computers, statistics was conceived as comprehending all aspects of collecting, analysing, interpreting, and presenting data. Its goal is to convert the large amount of data available into meaningful and useful information. It provides methodologies (such as hypothesis testing) to draw conclusions from data: these methodologies depend on well-defined parameters with well-studied properties.

On the other hand, machine learning (ML) refers to the algorithms that construct mathematical models that can be used to make decisions based on the data set under observation. ML algorithms are used for data clustering, classification, or prediction, where the decision process itself depends on the data and not the instructions inside the algorithm (or computer program). ML has become a fundamental tool in applied sciences. Many if not most state-of-the-art technological innovations now involve ML processes.

We may contrast statistics with ML as follows. In a conventional statistical data analysis, an algorithm (i.e. fixed procedure, usually implemented as a computer programme) is applied to the input data to compute characteristic parameters (such as mean, variance, etc.) according to definite mathematical formulas. On the other hand, a machine learning approach scans the data (often multiple times) and employs an optimization process to obtain model parameters which usually have no direct interpretation.

Artificial intelligence (AI) is the imitation of human intelligence in machines that are programmed to think like humans and mimic their actions and behaviour. It refers to tasks that require human-like intelligence, such as vision (computer vision) and perception, speech recognition, and language translation, prediction and classification. AI is an interdisciplinary science with multiple fields and approaches that interact with each other.

Big data refers to exceedingly large datasets (often measured in gigabytes or terabytes) that can be processed by computing to expose patterns, trends, and associations, especially relating to human behaviour and interactions. The size and complexity of big data means that none of the traditional data management tools can process it efficiently. Most important in big data are characteristics concerning its large volume, high throughput, and extreme variability.

The field of data science can be broadly construed as extracting value from data. It includes statistics, AI, ML, big data, and programming.

This book includes contributions from an international array of qualified researchers in the field who present and explain theoretical and practical models for artificial intelligence applications within data science with several applications. Each contribution was screened and pre-approved by all three editors, and then subjected to independent blind cross-review by two expert reviewers. Following are synopses of the different chapters.

Chapter 1 discusses AI for business data analytics. Business analytics (BA) is the extraction of intelligence from business data. BA involves organizing, processing, and examining business data for the purpose of gaining useful insights. The objective of business intelligence (BI) is to identify which datasets are beneficial and how they can be influenced to resolve problems and enhance the efficiency, productivity, and revenue. Although there is a fear that BI can lead to loss in employment, it continues to be a subject of research and be used in business. This chapter describes the different AI techniques being currently used in business intelligence. Two case studies in which the AI is used for business is described. Some open research problems in the areas of AI in business analytics are discussed. AI has several positive influences on how business is carried out and investment is done with respect to increasing sustainability. The chapter concludes that the adoption of AI and machine learning (ML) improves the overall decision-making within an organization by using ML insights to make intelligent and smart decisions.

Chapter 2 reviews the past achievements and future promises of digital transformation. The invasion of technology solutions in nearly every industry has sparked interest in emerging concepts in different sectors, with concepts such as big data, Internet of Things, virtual reality, artificial intelligence, and digital transformation dominating formal and informal discussions. Although this interest has recently extended from practitioners to scholars, the literature on major concepts such as digital transformation is still young and limited. This literature review was aimed at mapping out existing scholarly resources on digital transformation to provide a baseline understanding of the areas covered, defining the extents of adoption of technology, and understanding future trends in digitization. However, a deep rethinking and pragmatic steps of how an organization uses technology, people, and processes to fundamentally change business performance are referred to as digital transformation. It has been revealed that digital transformation has the potential to reshape the way business is conducted and how business models are practised.

Chapter 3 focuses on analysing relevant algorithms to develop a deep learning model to predict learner behaviour (learner interactions) in the learning process considering massive open online courses (MOOC). Various geographical, social, and learning behavioural features were used to build deep learning models based on three types of recurrent neural networks (RNN): simple RNNs, gated recurrent unit (GRU) RNNs, and long short-term memory (LSTM) RNNs. Results showed that simple RNNs gave the best performance and accuracy on the dataset. It is also observed the realization of some correlations between video viewing, quiz behaviour, and participation of the learner. It is found that one of the key benefits of this model is the fact that, by giving a good understanding of learner's behaviour, the model might guide teachers to provide personalized support and interventions to learners in the learning process. It becomes clear that simple RNN model could also be used to support adaptive content and learner pathways, by suggesting the revision or restructuring of the content and/or training path, closure of a course, or the launch of a new course. This model could also support data understanding by providing insight information, explanatory tools, and assisting in the understanding of the mechanisms underlying the desired outcome.

Chapter 4 explores and reviews that machine learning has been extensively used in the area of rainfall prediction. The studies are examined in terms of the source of the data, output objective, input features, pre-processing, model used, and the results. The review shows questionable aspects present in many studies. In particular, many studies lack a baseline predictor for comparison. Also, many references do not provide error bars for prediction errors, so that the significance of differences between prediction methods cannot be determined. In addition, some references utilize practices that permit data leakage, leading to overestimates of prediction accuracy. Different pre-processing like random shuffling used in the literature suggests that in some cases model performance is inaccurately represented. The aim of the survey is to make researchers aware of the different pitfalls that can lead to unreal model performance, which does not only apply for rainfall but for other time series data too. This chapter reveals that short-term studies typically rely on huge datasets and require deep learning applied to large feature sets to find hidden patterns in those datasets. On the other hand, long-term studies rely more on pre-processing methods such as feature selection, data imputation, and data balancing in order to make effective predictions.

Chapter 5 discusses cognitive computing, emotional intelligence, and artificial intelligence in healthcare. Cognitive computing entails creating computerized models to imitate the thought process of human beings while emotional intelligence encompasses being aware of and managing one's emotions, noticing and comprehending the emotions of others, and managing relations with others. Cognitive computing and artificial intelligence are trends being adopted in the industry. However, it is important to explore their impact on care delivery and the challenges associated with their use. In this chapter, an organized literature review involving seven studies was performed to evaluate emotional intelligence, artificial intelligence, and cognitive computing. Based on the findings, artificial intelligence and cognitive computing improve patient diagnosis and can accurately recommend treatments. In addition, cognitive computing and artificial intelligence enhance patient engagement and can help in streamlining administrative tasks. They can also help in reading patients' emotions and respond to them. They also possess the potential of comprehending emotions hence helping healthcare workers to offer patient-centred services. Some of the drawbacks noted with the use of these technologies include the risk to data privacy and security, job losses due to automation, and the lack of clarity concerning legal liability when machines are used to make decisions.

Chapter 6 explains the systematic review on application of data mining techniques in healthcare analytics and data-driven decisions. The review focused on analytical and theoretical of the application of data mining in healthcare analytics. The integration of healthcare analytics has continued to revolutionize the healthcare industry and it has helped in dealing with high readmission rates and medical fraud among other complicated issues in healthcare. The orderly review employed a preferred reporting items for systematic reviews and meta-analyses (PRISMA) technique to review research articles published on such applications of data mining in this area. The search process focused on healthcare analytics, data mining, artificial intelligence, and machine learning. The search results were filtered based on subject, year of publication, subject, peer-review status, and full-text availability with specific reference to open access journals. In analytics, data mining algorithms yield vastly improved accuracy and reliability over traditional data analysis techniques such as probabilistic forecasting. However, data mining algorithms have been successful in reducing waiting times and time to admission. These algorithms also prevent information loss and are useful in collecting and preparing data for research activities. On the decision support side, machine learning techniques have brought significant improvements in prognosis and diagnosis for a number of diseases and have contributed to personalized medicine. These algorithms also give valuable information on post-treatment reactions and responses to drugs. Theoretical papers in the recent literature discuss a variety of issues such as the role of healthcare analytics in disease control, data quality control, policies in healthcare, and patient privacy. It is also concluded that stakeholders are under pressure to integrate data mining algorithms into current healthcare systems.

Chapter 7 describes the design, implementation, and testing of a machine learning model that detects a malaria parasite *Plasmodium falciparum* using samples of thin blood smears on standard microscope slides that were taken by local laboratories in Khartoum, Sudan. A watershed segmentation technique is used to acquire the erythrocytes (RBC) from microscopic blood sample images, followed by a deep learning classifier. The classifier is based on a convolutional neural network (CNN) obtained using transfer learning. A base CNN is trained using a large publicly available data set of stained infected and uninfected cells; then several layers are appended to the CNN, which is retrained using locally obtained cell images, with image augmentation. Results showed that although good performance is obtained on the public dataset with the base CNN, the retrained CNN performs poorly on local images. Two reasons for this are the small number of local images available for training, as well as the poor quality of local images. It is recommended that further studies should consider improving these two factors to arrive at more reliable results.

Chapter 8 studies the automatic number plate recognition (ANPR) systems. ANPRs have the capability to automatically identify vehicles number plates from images and convert them to machine-readable ASCII characters. The systems have been developed using several algorithms and methodologies, including optical character recognition, convolutional neural network (CNN) or deep neural network, morphological operations, and edge detection. This chapter discusses a proposed automatic recognition system for Oman's number plates. For this purpose, this chapter presents a theoretical and analytical comparison between several previous works in this field to realize which algorithms are most suitable for plate recognition. A practical evaluation is conducted using actual number plates. According to this study, there are three main different levels of the recognition system: number plate detection, number plate recognition, and character recognition. The system is based on three key image processing algorithms: morphological operations for number plate detection, thresholding operations for number plate recognition, and CNN for character recognition. Several datasets were used to evaluate the accuracy and execution time of algorithms in the system. The execution time for recognition was estimated as sufficiently short to make the system suitable for real-time operation in realistic traffic conditions. Additionally, number plate overall extraction accuracy was 71.5%, while the CNN modelling gave between 96% and 100% accuracy on extracted characters.

Chapter 9 explores the social networks and presents how to use FastText model to detect real-time First Story on Twitter. Online social network (OSN) is microblogging that attracts millions of users on daily basis. Twitter is one of the microblog services which has millions of users. The own feature of Twitter is that it has 280 characters only for each tweet. Identifying the first tweet that happened in a specific time and place within a sequence of tweets means First Story Detection (FSD), since the story means a tweet. The term Frequency–Inverse Document Frequency or TF-IDF is a traditional algorithm that is commonly used for Text similarity applications like FSD. mTF-IDF means modified Term Frequency- Inverse Document Frequency used also in 2017. In this chapter, we made a comparison between using a model of machine learning called FastText and traditional methods such as TF-IDF and mTF-IDF to show which is more accurate using a pre-implemented open-source for FSD that uses storm platform which has some advantages like scalability, efficiency, and robustness in analysing the tweets in real time. Significant enhancements in the detection accuracy appeared in the results without any effect on the performance with FastText; our results show that FastText is better than mTF-IDF and TF-IDF in running time and accuracy.

Chapter 10 shows how musical notes are created using a bi-directional long short-term memory (LSTM) model with attention mechanism trained on MIDI data for generating unique music. Generating music is an interesting and challenging problem in the field of machine learning. Mimicking human creativity has been popular in recent years, especially in the field of computer vision and image processing. With the advent of generative adversarial networks (GANs), it is possible to generate new similar images, based on trained data. But this cannot be done for music similarly, as music has an extra-temporal dimension. So, it is necessary to understand how music is represented in digital form. When building models that perform this generative task, the learning and generation part is done in some high-level representation such as MIDI (musical instrument digital interface) or scores. The chapter proposes a model with attention mechanism capable of generating similar type of music based on MIDI data. The music generated by the model follows the theme/style of the music the model is trained on. Also, due to the nature of MIDI, the tempo, instrument, and other parameters can be defined and changed.

Liverpool, UK Mohamed Alloghani

Killeen, TX, USA Christopher Thron

Khartoum, Sudan Saad Subair

Acknowledgements

Writing a book is more rewarding than I could have ever imagined. I would like to take this opportunity to express my sincere gratitude to the UAE's government, leadership, and to His Highness Sheikh Mohammed bin Zayed Al Nahyan who devoted his life to raising the stature of the nation, ensuring our happiness, supporting and motivating us to limitless innovations, and providing everything that is useful for the advancement of the country.

I am indebted to so many, but I especially wish to acknowledge the encouragement, inspirations, and motivations of His Excellency Dr Sultan Ahmed Al Jaber.

This work would not have been possible without the support of Prof Christopher Thron and Prof Saad Subair. I am especially indebted to my parents and family, who have been supportive of my career goals and who worked actively to provide me with everything needed to pursue those goals. I am grateful to all of those with whom I have had the pleasure to work during this and other related projects.

Mohamed Alloghani

Contents

About the Editors

Mohamed Alloghani is an advisor (Research and Capacity Building) to the UAE's Minister of State for Artificial Intelligence, Digital Economy and Remote Work Applications at UAE Prime Minister's Office. Dr Alloghani is the first UAE national to graduate with a Doctor of Philosophy – PhD focused on artificial intelligence from Liverpool John Moores University. Dr Alloghani has 14 years of work experience in various public sectors such as oil and gas, judiciary/public prosecution, and healthcare. Due to his experience and work on strategic national missions, Dr Alloghani was delegated by the Minister of State for Artificial Intelligence and appointed by the Director General of UNESCO to represent the UAE in the Ad Hoc Expert Group (AHEG), in March 2020. The group comprised of 24 renowned specialists selected from around the world. He was the head of the UAE delegation G20 Digital Economy Task Force (DETF) in Feb 2020. Dr Alloghani's research interests focus on artificial intelligence, research and development, software engineering, and project management. He has authored and co-authored several scientific papers in regional, international, and refereed high-impact journals and conferences. He has published 51 peer-reviewed scientific papers.

Christopher Thron is Associate Professor of Mathematics at Texas A&M University-Central Texas. Previously, he was a communications systems engineer at NEC, Motorola, and Freescale Semiconductor. He has also held visiting positions at multiple universities in Cameroon, China, Nigeria, and Sudan. His teaching and research have been supported by grants from the US Fulbright Program, the International Mathematicians Union, and the U.S. Air Force. His research focuses on the application of computational mathematics and statistics (including machine learning) to various fields such as algorithm design, wireless communications systems, epidemiology, social sciences, and operations research. He has 9 patents granted and more than 40 publications in refereed journals. He has also presented conference talks and workshops in computational mathematics across the African continent.

Saad Subair was born in Sudan, on the banks of the river Nile, a few kilometres away from the capital Khartoum. He is Professor of Bioinformatics and Computer Science at the College of Computer Studies, International University of Africa (IUA), Khartoum, Sudan. Prof Subair obtained a BSc from the University of Khartoum; PGD, MSc (computer science), and PhD (bioinformatics) from UTM, Malaysia; and an MSc in genetics from UPM, Malaysia He is author and/or contributing author of several books, articles, and scientific papers published in the USA, Germany, Malaysia, India, and Arabia. He has been a keynote speaker at numerous regional conferences. Prof Subair is a member of scientific and academic committees at multiple universities in Sudan and Arabia. He is also member of high committees in the Ministry of Higher Education, Sudan. Prof Subair has trained hundreds of students in the fields of machine learning and bioinformatics, and has supervised and/or advised several research students who have achieved further successes in the UK and USA.

Chapter 1
Machine Learning for Business Analytics: Case Studies and Open Research Problems

K. Aditya Shastry, H. A. Sanjay, and V. Sushma

1.1 Introduction

Machine learning (ML) is the application of information technology and computers to imitate the intellectual capabilities of humans (Zahrani & Marghalani, 2018). ML can enable machines to solve research-based problems. The development of intelligence and reasoning capability in ML is still in the early stages (Trippi & Turban, 1992). Insights are created by the algorithms which are capable of successfully discovering patterns hidden in data. These patterns and insights learned by the algorithms provide the machines with the capability for making predictions and decisions in the future. Through the use of ML, computers may be programmed to perform tasks which ordinarily require human intelligence and reasoning, such as speech recognition and visual perception (Brynjolfsson & Mcafee, 2017). Naturally, to bestow upon computers, such complex capabilities require extensive programming and sophisticated algorithms.

One area in which ML has significantly impacted business is business intelligence (BI). BI refers to information that is gathered to provide insight into business decision-making. ML enhances the ability of BI to provide business enterprises with valuable insights leading to effective actions. For example, the trends in demand for products and services can be forecasted using ML algorithms, thus enhancing the profit-making capacity of business organizations (Zahrani & Marghalani, 2018).

K. A. Shastry (✉) · V. Sushma
Department of Information Science & Engineering, Nitte Meenakshi Institute of Technology, Yelahanka, Bengaluru, Karnataka, India
e-mail: adityashastry.k@nmit.ac.in; sushma.v@nmit.ac.in

H. A. Sanjay
M.S. Ramaiah Institute of Technology, Bengaluru, India
e-mail: sanjay.ha@msrit.edu

M. Alloghani et al. (eds.), *Artificial Intelligence for Data Science in Theory and Practice*, Studies in Computational Intelligence 1006,
https://doi.org/10.1007/978-3-030-92245-0_1

These algorithms have been increasingly employed by leading business companies to advise the management on policies and best practices that can result in increasing the company's profits. Even small and medium enterprises have also started to use ML for their businesses, due to the ease of access of ML technologies. Before the introduction of ML, business leaders were dependent on data from BI systems that were incomplete and inconsistent in nature. In contrast, ML software has the capability to analyze huge, complex datasets and offer valuable business insights which can assist business personnel in making improved decisions that are well-informed (Narula, 2019; Better Bayesian Learning, 2003). Organizations that use ML have competitive edge over other businesses since they can make effective decisions. However, it should be noted that ML business systems thus function as tools that support humans and are not substitutes for humans.

Deep learning (DL) is a subfield of ML that is developing means for providing business enterprises with real-time insights on how business is progressing. For example, humanoid robots in offices are operating more efficiently than humans (Better Bayesian Learning, 2003). These robots are fitted with ML tools for real-time data analysis, thus facilitating information retrieval. Unlike conventional business dashboards, robots can connect with other business systems so as to enhance coordination between business departments. DL has also found its way in business hiring in which DL and TensorFlow have automated the whole recruitment process by making the employers identify suitable candidates (Archana Bai, 2011). These DL applications enable the companies to locate and recruit appropriate candidates, thus enhancing productivity by reducing the downtime caused by the lack of skilled personnel.

The remainder of this chapter is organized as follows. Different techniques for business analytics are discussed in Sect. 2. Section 3 describes four significant real-world applications in the field of business analytics. Section 4 covers research areas that could be explored in business analytics. This is followed by conclusions in Sect. 5.

1.2 ML Techniques in Business

Nowadays, the availability of open-source ML tools such as TensorFlow and Sci-kit Learn has made way for organizations both small and large to effectively extract intelligence from data. Other examples include Amazon Web Services, the Google Cloud AI platform, and the Azure platform from Microsoft (Narula, 2019). This section discusses some ML algorithms that are used to improve business efficiency, as well as some sample use cases.

1.2.1 Bayesian Classifiers

Classification algorithms related to Bayesian theorems are simple and effective for many business applications. These rely on Bayes theorem, which is an important concept from conditional probability. It categorizes data based on the probabilities of specific attributes.

Two business use cases in which Bayesian classifiers can be utilized are outlined below (Narula, 2019):

- *Loan default prediction*: Suppose a loan officer in a bank needs to forecast whether an applicant of loan will repay the loan (non-defaulter) or not (defaulter) based on input attributes like annual income, loan amount, tenure of employment, equated monthly installment (EMI), etc. The output attribute would be the "past-default-status" of the applicant. The task would be to predict the binary class containing "yes" if the applicant is likely to default else no. Using this information, the bank could choose whether to grant loan for the applicant or not. The bank can also decide on the credit and interest rates to be given for each loan applicant.
- *Medical treatment recommendation*: Suppose a doctor wants to forecast whether the treatment given to the patient will be successful or not based on input features like levels of hemoglobin, blood sugar, blood pressure, current and previous medications, etc. The target attribute can be the "past-cure-status" of the patient. The predicted class may contain values such as "yes" indicating "prone to cure" or "no" denoting "not prone to cure." Effective treatment recommendations can be done through this type of prediction.

Bayesian classifiers may also be used for the classification of documents. For example, reference (Better Bayesian Learning, 2003) described a spam filter based on Bayesian classifier that was able to block 99.5% of spam mails without generating false positives (non-spam mails erroneously marked as spam mails). The email headers, contents, embedded JavaScript, and HTML were all utilized as inputs for the spam classification. Fifteen features are extracted from the input that indicated the presence of spam, on which the classification is based.

Some pros and cons of Bayesian classifiers are described below:

Pros (Martinez-Arroyo & Sucar, 2006):

- They perform better than complex algorithms in many cases while being easily implementable.
- The classification criteria can be understood without difficulty. As every feature is associated with a probability, the feature importance can be easily recognized.
- The classifiers are incremental, in that once a classifier is trained, it can be updated using new data without rescanning previous data.
- They are faster than most techniques, as the probabilities are computed beforehand, and classifications on new complex and large datasets can be made faster.

Cons (Al-Aidaroos et al., 2010):

- They assume independence of classifying features, which may lead to lower accuracy. For instance, consider a case where the words "online" and "medicine" only indicate spam when used in combination. A Bayesian classifier that used the two words as separate features would not recognize this interdependency, which would result in either false alarms or missed detections.

1.2.2 Decision Tree Algorithms

Decision tree (DT) algorithms are the easiest to interpret among all the ML algorithms. A series of if-then rules are generated by the DT algorithm, leading to the final classification of the data into predefined labels. DT falls in the category of supervised learning since the labels are predefined (Kotsiantis, 2013).

Following are some use cases in which the DT technique is utilized for BA:

- *Beverage classification:* An example is the classification of beverages based on features such as carbonated or not carbonated, leaves or beans, hot or cold. The DT classifier may classify the instance as iced tea depending on the features. Measures of impurity such as Gini index or entropy are utilized to determine the best split. Classification and regression trees (CART) are used in which every tree level is divided into two outcomes. The process repeats until further splitting is not possible or a certain threshold is reached (Zhong, 2016).
- *Subscription services:* Another application of DT is in predicting user signups for a subscription service (Jinguo & Chen, 2011). The classifier can be supplied with user data that records users' decisions on whether to sign for a free trial, test an interactive demo, and/or sign up for mailing list and how they found the website (via search engine, friends, social media, etc.), as well as their decisions on whether to convert to paid service. Based on this data, the DT constructs a tree that not only helps to predict which users are likely to become paying customers but also can demonstrate steps frequently taken in the conversion of a trial user to paid user (e.g., social media → mailing list → paid subscription). This can provide valuable insights to the company for targeting customers.

Some pros and cons of decision trees are listed below:

Pros (Al Hamad & Zeki, 2018):

- They are easily interpretable even by non-technical users, as the trees resemble flowcharts with "if-then" branches and leaf nodes that signify the classes.
- They are suitable for both numerical and categorical datasets and can provide meaningful information even for small datasets.
- They consider dependencies or interactions between input variables. For instance, for the "online medicine" spam example above, DT can label "online" and "medicine" occurring separately as non-spam while identifying joint occurrences as spam.

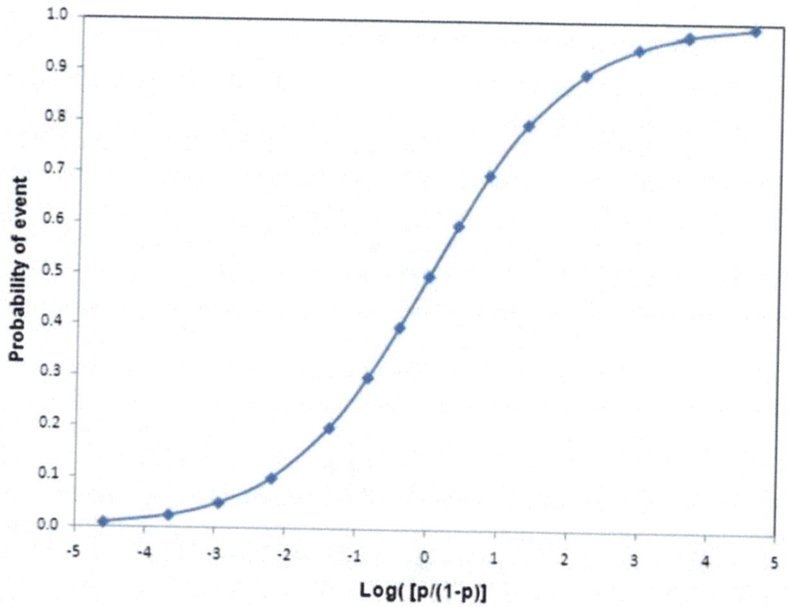

Fig. 1.1 Graph of logistic function. The input value is interpreted as "log odds" (log of the ratio of binary class probabilities), and the output is the probability of the first binary class

Cons (Al Hamad & Zeki, 2018):

- Overfitting of data may occur where the model predicts accurately on training data but fails on test data (unknown data). Pruning can be utilized to enhance the accuracy of the DT models.
- Trees may become large and complex for certain datasets, which can lead to slower classification.

1.2.3 Logistic Regression

Logistic regression (LR) is a modification of conventional regression which can be used for binary classification, i.e., predicting whether or not the inputs correspond to an instance of a given class (Hui-lin & Feng, 2011). A LR model first computes a weighted sum of numerical inputs to produce a single value, which is then converted to a number between 0 and 1 by applying the S-shaped logistic function as shown in Fig. 1.1.

A threshold value (also known as decision boundary) for the output is set by the user to determine whether or not the given set of inputs is assigned to the specified class. For example, logistic regression may be applied to features of business transactions to predict whether or not the transaction is fraudulent. If the LR output applied to the features of a given transaction assumes a value greater than the user-defined threshold, then the transaction is classified as a fraud and otherwise as legitimate (Haifley, 2002).

LR models are determined by the weights assigned to the different input features. These weights so as to minimize a cost function (which represents the model error) when the model is applied to training data are utilized by LR to compute the error level for each prediction when compared to the training record. This minimization may be accomplished through a process of gradient descent, during which the weights are successively adjusted. Once the LR is fully trained, it can be used on new data to make practical predictions (Hui-lin & Feng, 2011).

LR can also be used for multiclass problems in which inputs are to be assigned to more than two classes. This version of LR is termed "multinomial LR" and can be implemented by performing multiple binary classifications. For instance, suppose documents in a repository are to be assigned among three classes: health, politics, and economics. Under the "one versus all" procedure for multinomial LR, three binary LR models are constructed: one for "health" or "not-health," the second time for "politics" or "not-politics," and the third for "economics" or "not-economics." The three binary models are run on input features derived from each document, and the results are combined to assign the document to a single class.

Some examples of the use of LR in BA are given as follows:

- *Credit scoring* A Finance Organization called ID Finance (idfinance.com) develops forecast techniques for credit scoring. They require their scoring techniques to be clearly understandable by their consumers. In such a scenario, LR becomes a good technique since it can determine the attributes (i.e., features) that have the biggest influence of forecast outcomes. Using LR, an optimal number of attributes can be determined, and unnecessary features can be eliminated.
- Websites that provide online booking can employ LR to forecast the intentions of their consumers. Forecasts may include where and when the user will travel, whether the user will modify his/her plan, preferred routes and intermediate stops, preferred hotels, etc. Since these attributes are categorical, multinomial LR is a suitable ML algorithm.

Other practical applications of LR include sheet metal analysis, forecasting safety concerns in coal mines, and several applications in healthcare (Bhattacharyya & Bandyopadhyay, 2014).

Pros and cons of LR may be described as follows (Zekic-Susac et al., 2004):

Pros:

- It is relatively easy to understand the LR model output, compared to other models. Since the output of LR is between 0 and 1, it can be interpreted as probability. For example, 0.276 can be interpreted as 27.6% possibility of a credit card transaction being fraud. The weights assigned to features denote its importance in classifying the target.
- LR can be adapted to use with categorical data, by making use of dummy variables. For example, "blue or not blue" can be converted to "1 or 0" where blue – 1 and not blue – 0.

Cons:

- If the number of records is small compared to the number of attributes, LR may overfit.
- LR is only suitable for a limited class of situations in which a linear decision surface can separate the alternatives.

1.2.4 Support Vector Machines

Support vector machines (SVMs) form a complex and powerful classification technique, originally designed for binary classification. It identifies a maximum margin hyperplane that divides the data most accurately. In some cases, if linear separation is not possible, the data is transformed to another space using kernel functions in which the data is linearly separable. For example, using the squares of GPA and SAT scores instead of the original scores may make it easier to separate between two classes of students (Lovell & Walder, 2006).

The construction of SVM classifiers based on training data involves detailed technical knowledge and may be challenging even for those who are mathematically proficient. However, once constructed SVM are easy to use. Recently, a variant of SVM called support vector regression (SVR) is being utilized for prediction (regression) tasks (Yu et al., 2013).

Some practical use cases of SVR are as follows:

- *Handwriting recognition*: In real-world practice, SVMs are extensively utilized in handwritten recognition, identification of facial expressions, and image classification. For instance, the numbers in Persian/Arabic scripts were identified with 94% accuracy by the SVM. There were ten classes corresponding to 0 to 9 digits. The classification was performed digit by digit into one of these classes. In this application, the SVM was found to be more accurate than neural networks (Parseh et al., 2020).
- *Credit evaluation of business websites using SVM*. In (Guo-sheng & Guo-hong, 2007; Goh & Lee, 2019), authors have used SVM to evaluate the trustworthiness of the business websites. Authors have proposed a weighted SVM to score credit ranks of business websites. Different SVM kernels such as polynomial kernel, RBF kernel, and sigmoid kernel have been tested.

Pros and cons of SVMs are listed below (Karamizadeh et al., 2014):

Pros:

- Like LR, the SVM technique can be used for multiclass classification by using multiple "one versus all" SVM classifiers.
- SVM can also be modified for mixed attribute sets comprising of numerical and categorical data.
- They perform well for high dimensional data.
- They are highly accurate in many applications, if provided with suitable training data.

Cons:

- SVM is a black box method in that its classification criteria are not easily understandable by the user. For example, the DT and Bayesian classifiers can be easily traced using pencil and paper. In contrast, the SVM involves complex mathematical transformations making it difficult to trace the classification even for a technical person. The reason for a specific SVM classification may be ambiguous or unclear.
- Unlike DTs, SVMs usually need a very large dataset for training as it cannot infer intelligence from small datasets.

1.2.5 Artificial Neural Networks

The concept of artificial neural networks (ANN) was originally inspired by the structure and functioning of the human brain. The human brain consists of neurons for transmitting information to and from other interconnected neurons via electrochemical signals. Similarly, ANNs consist of mathematical elements called "neurons" which have multiple inputs from and outputs to other neurons. Note that ANNs are also referred to as neural networks or neural nets (Mishra & Srivastava, 2014).

The multilayer perceptron (MLP) is a type of ANN containing at least three layers of neurons consisting of the input layer, one or more hidden layers, and the output layer. The inputs to neurons in the input layer consist of the features of the input dataset. In each neuron, an "activation function" is applied to the neuron's input to produce an output, which is then passed to one or more neurons in the first hidden layer via connections known as "synapses," which are assigned different weights. Each neuron in the hidden layer then forms the weighted combination of its own inputs and applies its own activation function to produce an output, which is then passed to multiple neurons in the next layer via a second set of weighted connections. This process is repeated by all subsequent layers, up to the output layer which outputs the ANN's estimate of the feature(s) or attribute(s) to be predicted (Gupta et al., 2019).

ANNs must be trained using input data that has previously been classified and labeled. This data is used to incrementally adjust the ANN weights through a process called "back propagation." For example, if a training image labeled as "lion" is misclassified as "tiger," then the back propagation algorithm updates the weights of the synapses towards producing the correct output. After multiple rounds of training, the synapse weights eventually converge so that the ANN correctly identifies most of the training images (Abiodun et al., 2018).

Deep learning refers to neural networks with multiple hidden layers with structured interconnections. Besides layers of neurons, deep ANN's contain other types of layers for sampling or recombining neurons' outputs. Deep learning ANN's may be applied to highly complex image and audio datasets (Vasundhara & Seetha, 2016).

Several applications of ANNs can be found in the real world. Amazon makes use of ANNs for generating recommendations for certain products. Massachusetts General Hospital uses deep learning to improve the diagnosis of patients so that better treatments can be provided. Facial recognition is performed by Facebook using ANNs. The Google Translate website also uses ANNs for automatic translation (Alam, 2019).

Pros and cons of ANNs may be described as follows (Xie et al., 2011):

Pros:

- ANNs are scalable. Their performance tends to improve with massive datasets in contrast to other algorithms whose accuracy decreases with increase in dataset size.
- The training of ANNs can be done incrementally as new data are added. Hence ANNs may be kept online continuously without the need for off-line training.
- ANNs may be stored efficiently, as only the synapse weights can be provided. Similar to the Bayesian Classifiers, ANNs are denoted by a list of numbers representing the weights of the connections. In Bayesian classifiers, the probabilities of features represent the numbers.

Cons:

- ANN is a black box method, and even experts cannot interpret an ANN's basis for its decisions. The problem is exacerbated by the fact that ANNs may contain thousands of nodes, where each node may be connected to dozens of synapses.
- ANNs require very large datasets for training, even larger than those required by SVMs.
- ANNs must be customized for each individual application. Although software packages for constructing ANNs are available, skilled programmers are required to construct an ANN that are suitable for a given task.

1.3 Machine Learning Applications in Business

This section describes in detail several recent real-world applications of ML in business.

1.3.1 *Application-1: Development of a Repair Intelligence Platform Using Sensor Data, OEM Manuals, and Repair Invoices*

The AI software company Predii (www.predii.com) collaborated with the tool manufacturer Snap-on Incorporated (www.snapon.com) to create an integrated repair intelligence platform (Faggella, 2019). The platform is designed to facilitate repairs by more effective use of various forms of vehicle data, as described below.

First, Predii systemized and facilitated the process of accessing vehicle documentation. Exhaustive product manuals are produced by OEMs of vehicles for model types such as different models of trucks, cars, etc. These manuals consist of thousands of pages containing detailed specifications of vehicles, troubleshooting, and diagnostic procedures. Predii constructed a query-able database to enable technicians to effectively utilize these lengthy manuals. Natural language processing (NLP) was applied to vehicle manuals' contents to produce a well-structured database which automatically categorized the key concepts from each manual. This made searching through the manuals easier by creating a foundation for contextual searching along with extraction. For instance, a technician who searches for the phrase "Brake oil Ford" will obtain information related to brake oil details for all the Ford vehicle models. This application is a good example of unsupervised ML, in which the classification algorithm is able to learn on its own, without any prior training or supervision.

Second, Predii software synthesized and interpreted diagnostic information from vehicle sensor data. Many modern vehicles are installed with sensors that track tire pressure, temperature, speed, rotations per minute, etc. These sensors monitor the operating conditions of the vehicle and enable technicians to diagnose issues related to maintenance during breakdowns or failures. For instance, if a car is facing internal heating issues, then heat sensor data can provide valuable insights for solving the issue. The software built by Predii is able to analyze this data and identify potential future failures, thus leading to better preventive maintenance of vehicles. Figure 1.2 shows the user interface for accessing sensor data information. Items marked in red are the potential anomalies that the ML algorithm has identified from the telemetry sensor data, based on comparisons with historical data. Prompt actions for maintenance are also proposed by the software.

Finally, the repair intelligence platform assembles and interprets service order data originating from the repair shops. In the automotive industry, repairs and maintenance are performed by third-party vendors, and invoices are created by different technicians who work on the same vehicles independently. Combined information from invoices and warranty data can be used to provide technicians with valuable insights for efficient diagnostics. For instance, the most common type of repairs done on specific vehicle models can be found using ML applied to repair order data. Predii was able to construct a workable database from the unstructured repair order data, together with a platform for analyzing the unstructured repair

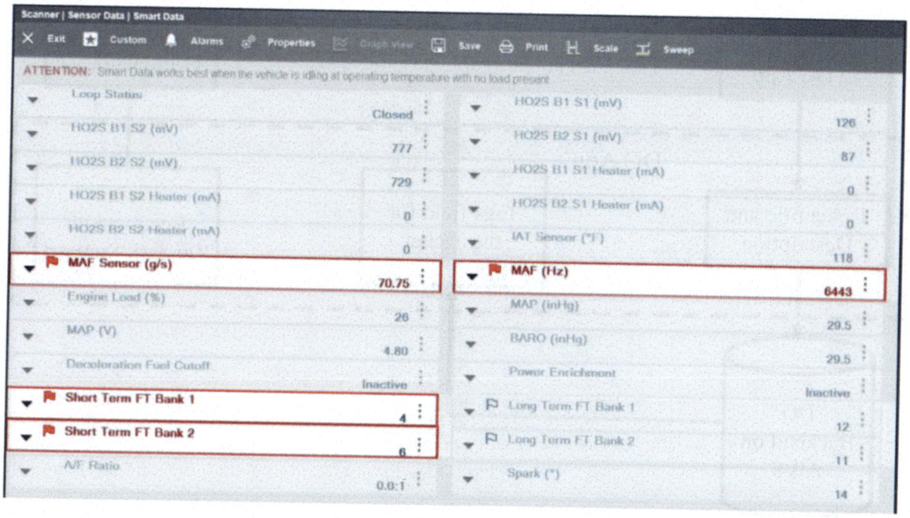

Fig. 1.2 Predii's repair intelligence platform (Faggella, 2019) for interpreting sensor data

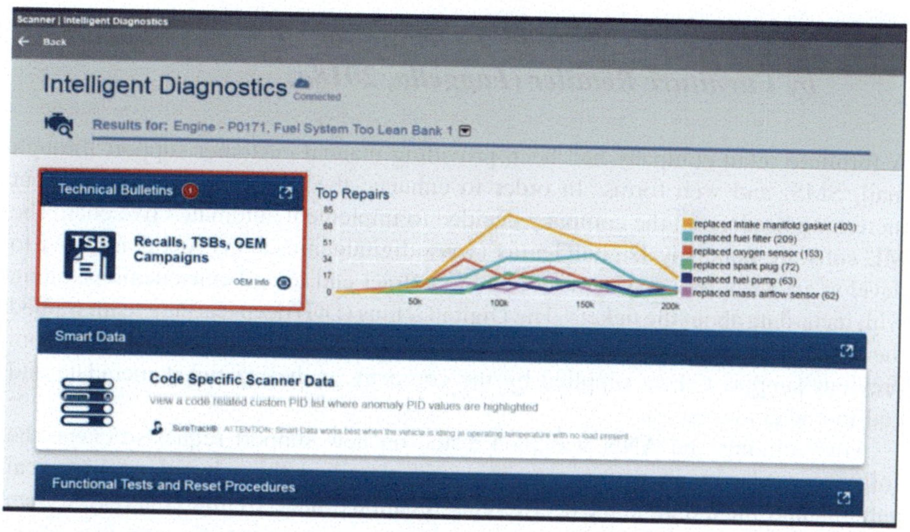

Fig. 1.3 Predii's intelligent diagnostics interface for analyzing repair order data

order data to extract diagnostic patterns and relationships. Figure 1.3 depicts a snapshot of the Intelligent Diagnostics screen showing the distribution of past repairs associated with a particular diagnostic condition (fuel system too lean).

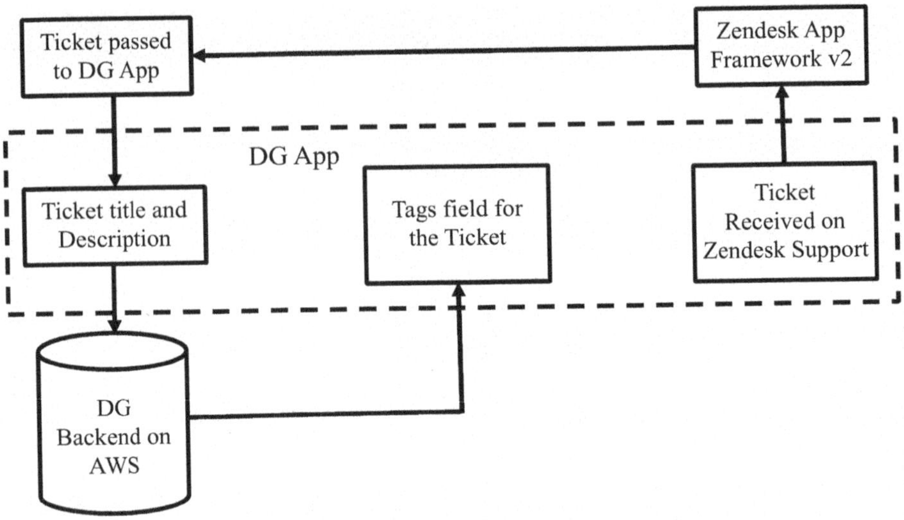

Fig. 1.4 A visualization of the automation process behind DigitalGenius (Faggella, 2018a)

1.3.2 Application-2: Machine Learning for Customer Support by Furniture Retailer (Faggella, 2018a)

A furniture retail company had been providing manual customer support through mail, SMS, and web forms. In order to enhance the quality of support without increasing headcount, the company decided to implement automated live chat. The ML software company DigitalGenius (www.digitalgenius.com) was contracted to develop an ANN-based system, devised to extract end to end conversations, along with meta-data about the tickets. The DigitalGenius (DG) deep learning team trained the ANN in secured instances of Amazon Web Services (AWS) using data from previous support tickets supplied by the company, including ticket metadata and end-to-end conversations.

After training, the ANN was then tested on new support requests/tickets: the software read new incoming messages and classified them based on historical patterns and meta-data context. The classifications made by the ANN were then validated by the customer service team. After proper testing, the app was made available to all customer agents.

Figure 1.4 gives a flowchart showing the integration of ML into customer support. The customer tickets are received on the customer service platform Zendesk (www.zendesk.com), which passes the information to the DG ML software, which in turn processes the information and assign appropriate tags in order to properly classify and route the incoming requests and facilitate rapid and appropriate agent response. Apart from forecasting and tagging the incoming tickets, the ML software also provides recommendations to the agents for ticket responses. Previously, the entire process was performed by human agents: currently, about 70% of the incoming tickets are being handled by the ML software using automated tags. The ML is able to quickly handle the tickets with less waiting time, while agents are able to spend their time in more useful work such as researching and improving the macros.

1.3.3 Application-3: Business Intelligence Apps Built on Machine Learning (6 examples of AI in business intelligence applications, n.d.)

Certain organizations have expertise in developing general purpose applications. This section discusses the various business intelligence apps created with ML. The use of ML in four business intelligence apps has been examined. The first application discussed is the SAP Software Solutions. The second application which is discussed is Domo whose domain is development of dashboards for business. The third application which is elaborated is the Apptus app which specializes in establishing valuable links between what the customer expects and the revenue realization by the organization. The Fourth application is Avanade that uses ML for generating valuable insights into business.

1.3.3.1 SAP Software Solutions: Machine Learning for Turning Databases into Useful Intel

SAP Software Solutions (www.sap.com) is a company that develops enterprise resource planning (ERP) solutions. ERP comprises core areas of business such as procurement, production, management of materials, finance, marketing, and human resources. One of the SAP's platforms is the High-Performance Analytic Appliance (HANA) which is used to manage the information in the collected databases. That is, structured data such as the information of customers or transactions related to sales are ingested from apps, databases, and other sources by HANA. This data may be collected from several business access points like desktop PCs, mobiles, financial transactions, production plant equipment, and sensors. Smartphones or tablets can be used by company staff to record purchase orders. HANA analyzes the data from these and can spot trends and irregularities. For example, HANA can notify of unusual customer orders or whether assembly-line equipment that is running slower than normal. One of HANA's distinctive advantages is the ability to handle high volumes of transaction records quickly, due to the fact that it stores frequently accessed data in RAM rather than on disk. This paves the way for real-time accessing of the data that can be effectively utilized by the applications and analytics that are developed on top of the HANA platform, thus enabling rapid managerial response to changing conditions.

HANA customers include large enterprises such as Walmart and GM, as well as mid-size and smaller companies. Benefits reaped include reduction in the cost of infrastructure and increased operational efficiency. According to an SAP report, ten enterprises that utilized HANA projected an average annual benefit of 19.27 million dollar per enterprise compared to the average yearly investment of 2.41 million dollars over a period of 5 years.

1.3.3.2 Domo: Machine Learning for Business Dashboards

Besides companies like SAP, smaller firms are also active in the development of ML platforms. The software company Domo (www.domo.com) represents a rapidly growing organization that has raised more than 500 million dollars in funding. Domo offers a cloud-based dashboard that aids organizations in making decisions. This dashboard is scalable and can be used by small teams of 50 or by bigger companies. Currently, 400 software connectors allow Domo to gather information from third-party apps that can be utilized to deliver useful insights for business. This permits organizations using Domo to gather data from popular applications such as Facebook, Salesforce, Shopify, Square, etc. for gaining insight about their customers, inventory, and sales. For example, the end users of Domo (who are usually merchants) can extract information from their e-commerce software and Shopify. This leads to effective management of online stores. The information extracted can be to generate reports and to identify real-time trends like performance of products that can be shared to any device utilized by the organization.

A set of novel features called Mr. Roboto was introduced by Domo that used AI, ML, and predictive analytics. Mr. Roboto is able to deliver recommendations and valuable insights for decision-makers. Whenever any anomalies and/or new data patterns are found, the platform generates alerts and notifications similar to those utilized in cybersecurity. The early detection can aid the predictive analytics and assist organizations to forecast the ROI for real-time marketing and make other predictions related to sales.

Currently, large organizations such as eBay and MasterCard are utilizing the Roboto platform. The media company Univision (https://www.univision.com/) uses the Domo platform with connections to applications like Google Analytics, Adobe Analytics, and Facebook in order to obtain high value for its programmatic advertising. Around 80% growth was achieved by Univision after using Domo.

1.3.3.3 Apptus: Machine Learning in Sales Enablement

Apptus software (www.aptus.com) specializes in establishing valuable links between what the customer expects and the revenue realization by the organization. The eSales solution of Apptus automates merchandising based on the forecasting the reactions of consumers. The software has combined big data and ML for determining the products that could be recommended to the user when he/she is purchasing online. When a customer makes purchase online on sites that use Apptus eSales and starts typing product items, then the ML solution can forecast and automatically display the associated search phrases. Products associated with the search terms will also be displayed.

One client of Apptus is Bokus.com, a Swedish online book retailer. Bokus.com was able to reduce the unnecessary revenue overheads by using Apptus that made the conversion of site visitors to customers seamless. Bokus reported an increase of 100% in its customer conversion rate per online recommendation newsletter.

1.3.3.4 Avanade: Machine Learning for Business Insights

Microsoft and Accenture have formed a joint venture called Avanade (www.avanade.com) that leverages the Cortana Intelligence Suite and other predictive analytics to provide data-based insights. Pacific Specialty which is an insurance company used Avanade in developing an analytics platform for providing its staff members insight into business. The objective of this exercise was to utilize the data of customers and their policies in generating more growth and profit. By analyzing the behavior of the policyholder and the trends via analytics, the idea was to better advise the new product developments.

1.3.4 Application-4: Provider of Services for Smart Home Using Machine Learning (Faggella, 2018b)

digitalSTROM AG (DS) is a German company that offers connectivity for smart homes by integrating electrical appliances into a network of hardware and software. Customer's data such as energy and water consumption, status information (working/not working, in/out, etc.), and monitoring of short-term tasks (whether or not someone rang the doorbell) are also tracked in a DS-equipped smart home. The company was faced with the challenge of representing this information to consumers in an easily comprehensible manner. Previously, the company used charts that were difficult for end users to understand.

To address this problem, DS employed an NLP platform called Automated Insights Wordsmith (AIW) that presents the consumer information in the form of customized reports in printed and audio formats. These reports are received by the consumers via an app. Using this app, the consumers are able to obtain specific smart home information upon request. An example of the process of requesting of information by a consumer is given in Fig. 1.5.

As shown in Fig. 1.5, the steps are given below:

- Consumers use the DS app installed on their mobiles to make queries: for example, they may ask how much electric power was utilized on a specific day.
- The app then collects the information pertinent to the query.

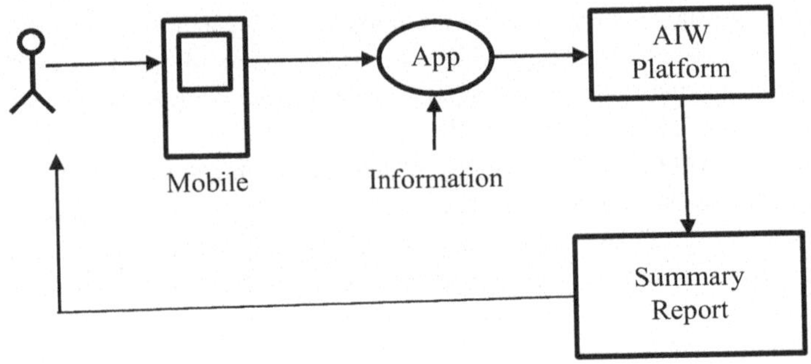

Fig. 1.5 Application scenario of AIW platform usage by DS company (Faggella, 2018b)

- It sends this gathered information to the AIW platform which then produces a summary report which is read out aloud/displayed to the consumer.

The key point to consider from this case study is that these types of applications related to natural language processing (NLP) can be utilized for personalizing the information of consumers and knowledge-based queries systems that possess linked infrastructure. Some of the applications being implemented in other business scenarios are listed below:

- User manuals of system for home applications can be customized based on individual preferences such as the techniques of troubleshooting already tested, the devices and applications of house, the type of mobile used by the consumer, location of the user, etc. Basically, information which is precise and contextual will be provided by creating manuals specifically for every consumer.
- A provider of marketing services may construct written summaries which are human-like for specifying the progress for campaign for marketing. The software may also hint at certain actions that can be performed by the team of marketing experts in the near future.
- A summary of assets specified in natural language may be created by a manufacturing industry for anomaly description in the data of sensors received from its devices. Manager suggestions for resolving the potential challenges may also be incorporated.

1.3.5 Application-5: Machine Learning Powered Customer Sentiment Analysis by Microsoft (Faggella, 2018c)

This case study discusses about how Microsoft's Customer Market Research (CMR) adapted ML technology to extract useful insights about the consumer surveys received. The CMR team formerly spent excessive time and resources on the design, deployment, and analysis of the surveys of consumers for extracting useful data for the organization. To resolve this issue, the CMR team obtained social media data from which it attempted to extract useful knowledge. This was planned in order to lessen the work done on manual survey collection and on improving the marketing research.

The CMR team of Microsoft framed flexible questions like "What do consumers have to say about the style and color of the new product?". For sorting the vast storage of comments on social media, the CMR team employed the text analysis engine created by Lexalytics (www.lexalytics.com) to analyze content related to brands, interest themes, and products. Data warehouses were utilized for storing the "feed" data from social media. The Lexalytics engine processed this social media data for determining the trends and perceptions out of the text data stored inside.

The feelings of the consumers were analyzed and visualized using Semantria Storage and Visualization software (SSV), also produced by Lexalytics. A team from Microsoft verified the SSV analysis using survey information. Thanks to this ML technology, Microsoft has been able in some situations to replace slow, expensive surveys with faster, cheaper social feedback, as well as to identify gaps in existing surveys.

Like Microsoft, other big consumer organizations are employing ML for getting faster consumer feedback on their products instead of relying on consumer surveys. It is not always possible to get the direct feedback of consumers on products and brands. Most consumers give no feedback since there is no direct benefit to them. Using ML, the feedback of consumers may be extracted from their tweets or opinions that they publish in social media. This better feedback aids organizations in improving their products as per the opinions of their consumers.

In the future, as advanced ML technologies are developed, the task of monitoring the media will become more prevalent for mining useful information from consumers. Some of the scenarios in which ML can be employed for performing this are listed below:

- Sentiments of consumers related to hygiene and sanitation can be extracted by hotel chains using ML.
- Opinions associated with a novel model of car may be gathered using ML by the manufacturers of cars when advertising through TV/online videos.
- A business firm selling candies may want to identify patterns among the consumer sentiments, their products, and their actual product sale. This may assist the firm in forecasting demands of their products in regional areas.

1.4 Research Areas Related to Machine Learning for Business Data Analytics

This section describes some directions of active research in areas related to ML for business data analytics.

1.4.1 *Research Directions in Big Data Analytics with Hadoop (Lim et al., 2013)*

The Hadoop software library (hadoop.apache.org) from the Apache Software Foundation provides a java-based framework for distributed processing of data intensive applications. It has been extensively applied for the analysis of complex datasets involving several record comparisons and active movement of data among servers. Hadoop has become an essential tool in big data analytics, which is often used together with various software platforms designed for data collection, distributed data storage, and data mining (Henschen, 2011). The top 3 commercial suppliers of databases (IBM, Oracle, and Microsoft) have all adopted Hadoop.

The processing of big data is done using Hadoop that involves summary statistics. With regard to Business Intelligence (BI) involving unstructured data, new advanced analytics of text, indexing of images, and one-time processing needs to be researched and developed in the MapReduce or distributed Hadoop environments. Hadoop MapReduce is a software framework for developing applications that process large quantities of information parallelly on huge clusters (thousands of nodes) of product hardware in a trustworthy, robust fashion. Graphical representations of big data are commonly used. In these representations, the nodes denote users or items; edges signify social relationships, flows of information, and adoptions of products. Graph mining is conducted in order to obtain insights about behavior of consumers in the graphs. Graph mining with the MapReduce or Hadoop Framework is a recent research topic which has not yet been thoroughly explored. One particular subtopic is the calculation of graph statistics such as the diameter of the graphs derived from big data. Along these lines, Kang et al. (Kang et al., 2011) built a Hadoop diameter and radii estimator (HADI).

One important practical challenge is the selection of optimal MapReduce/Hadoop frameworks for different applications in business analytics. Solving this issue requires understanding the constraints and strengths of frameworks for implementing sophisticated algorithms. Partitioning of analytics operations between the Hadoop framework and other frameworks should be considered: for example, it may be possible to use Hadoop for preprocessing, while the remaining operations are performed elsewhere.

The migration of legacy systems to a big data framework is another important challenge. This type of framework shift requires major changes in the existing infrastructure since the legacy data must be moved from traditional database servers to non-structured databases which do not conform to standard relational database standards. New indexing schemes must be generated, and existing applications must be reimplemented. Developers who are skilled in the domain of BI are required. It is important for businesses to have accurate forecasts of potential costs, since the cost of migrating to big data framework may be huge.

1.4.2 Research Directions in Text Analytics

Systems for information retrieval (IR) such as search engines have evolved into complex systems comprising fast algorithms that facilitate sophisticated indexed searching. Text-based business enterprise systems have incorporated these search engines for document management purposes. For large-scale databases or databases whose contents are dynamic, methods like in-memory processing and real-time processing are utilized. Semantic searching in another research aspect involves text analytics in the context of semantic matches as well as measures of semantic transfers (Agarwal et al., 2006; Guha et al., 2003).

The domain of NLP has seen significant advancements during the past decade from extracting information to the development of automatic answering systems. These systems are constructed by integrating the big data technology for training purpose and statistical NLP for developing language models. The NLP has been successfully applied not only for representing text like phrases, relationships, bag of words, and entities but also in Q/A (Question Answering) systems and machine translation and for identifying the topics and events. IBM Watson represents a sophisticated Q/A system that is devised by utilizing advanced analytics in order to interpret the context and meaning of human language. Some of the areas in which Q/A systems find their use are in defense, education, and health (IBM, 2011).

Another significant direction of research in information extraction and search engines is how to represent the search queries by considering the different preferences of the user. Apart from the page-rank algorithm used by Google (Page et al., 1999), researchers are developing other schemes of query representation. Product reviews published online are one example of how query results may be displayed in a user-friendly fashion. Compact search outcomes that are rich in information and less redundant need to be generated for providing advertisements and mobile-related web queries.

Opinion mining is another important research area. From online text, useful sentiments can be extracted from which useful insights may be obtained by the stakeholders. The development of social media contents and Web 2.0 has enabled better understanding of the opinions of the general public and consumers about social events, company strategies, political activities, campaigns related to marketing, and preferences of products (Abbasi & Chen, 2008; Chen & Zimbra, 2010).

Text analytics also faces major challenges in obtaining useful information from web content that may contain grammatical and spelling mistakes, emoticons, abbreviations, informal slang, mixed languages, etc. In such cases, the traditional or standard text analytics does not work accurately. Additional challenges are posed due to short forms of messages used frequently in SMS, Twitter, and other social media platforms (Lin et al., 2011; Savage, 2012). To resolve or address these issues, innovative approaches are needed.

The difficulties of text analytics are compounded when online sensors/applications generate continuous stream data that must be analyzed in real time by BI applications. Significant amount of research work has been carried out in analyzing the data streams that are structured, but additional research is required for the analysis of data streams which are unstructured in nature (Gaber et al., 2005). As the text data from web and social media is growing rapidly, future research must focus on analyzing these unstructured data streams to gain insight of the underlying content.

1.4.3 Research Directions in Network Analytics

In many business organizations, the data related to different transactions or different customers are considered as independent. This paradigm is limited, as the records or data may have useful connections, either directly or indirectly. For example, in sites related to social networking, the users are related to through variety of links such as friendships, shared communities, trusts, interactions through messages, etc. As another example, in online auctions, the buyer is connected to various sellers, and the purchase transaction for an item is linked to previous bidding transactions. Yet another example is afforded by telecommunications providers, who may analyze the phone calls between different customers to derive valuable insights about their preferences. All of these cases involve links between data items, so that the dataset may be represented as a graph or network with connections. From the network's structure, useful insights about customers' behavior can be extracted. By knowing customers' preferences, customers can be offered better products and services, and customer churning can be avoided.

Several research directions within the area of network analytics may be identified, as detailed in the following subsections.

1.4.3.1 Link Mining

In link mining, the links between nodes of a network are discovered using ML. As described above, these links may signify social relationships, exchanges of email, adoptions of products, or customer representation. Network data may be incomplete and possess missing links, which may hamper the ML from deriving useful patterns. Hence, sophisticated techniques are required for inserting missing links and predicting future links. Reference (Liben-Nowell & Kleinberg, 2007) utilized the information related to topology for performing link mining. The future or missing links can be predicted using popular techniques like Jaccard's coefficient, Katz measure, common neighbors, and the Adamic/Adar measure. These measures presume that the nodes comprising high topological proximity between them are more probable to possess links or relationships among them. However, the techniques related to topology do not perform well when new nodes join the network. When the link and node attributes are considered, the accuracy of link mining may be improved. If the topology and attribute approaches are considered during link mining, then links to new nodes can be accurately predicted.

1.4.3.2 Community Detection

Community Detection (CD): Communities of users form a network for various reasons such as relationships among family, product adoption patterns, or friendships. These relationships may lead to dense clusters inside a network. Since communities are formed of people having similar preferences, community detection can uncover common preferences and focus on the similar patterns. With regard to banking applications, communities of users may denote different segments of customers. It is, therefore, crucial to design variety of service and product packages that target certain segments of customers. Recently, the detection of communities has formed a very active area of research. Various survey papers on CD have been published (Fortunato, 2010; Porter et al., 2009). When networks are represented as graphs, graph partitioning methods can be applied to identify minimal cuts. This leads to the discovery of dense subgraphs that signify communities of users.

The main objective is to partition the network into several subgraphs so that minimal links are generated between the subgraphs. Modularity is a goodness measure that has been utilized to evaluate the partitioning accuracy of a network. For each network link, the measure of centrality like betweenness is estimated. The links having high values of centrality are eliminated in each iteration until a network with good partition is obtained. The resulting network should possess optimal goodness value.

1.4.3.3 Social Recommendation

As consumers spend more time online, retail industries are going digital in developed countries. Significant e-commerce organizations like eBay, Flipkart, and Amazon sell their products online. Geographical boundaries do not apply during online shopping, so the customers have a variety of purchase options on the web. Since customers are distributed globally, the recommendation of suitable products to end users has become more difficult.

Product recommendations are typically performed using a technique called collaborative filtering (Herlocker et al., 1999). This type of filtering performs accurately if target users have bought products previously. However, for new target users, collaborative filtering has no information on which to base recommendations. This is called the cold-start problem. To resolve this problem, network data about consumers may be utilized, and users linked with the target user can be mined. This requires the integration of the social aspect with collaborative filtering to utilize purchase information from users linked with the target user to make suitable recommendations, based on the assumption that linked users will have similar preferences. This is called social recommendation, which is a novel research topic.

New social recommendation methods that have introduced social factors are being devised. For instance, extended factorization of matrices with weights of social links for forecasting the rating was developed by (Ma et al. (2008)). The latent Dirichlet allocation (LDA) model has been applied by (Chua et al. (2011)) to find the social correlations among users to enhance the predictions of the adoptions of items.

But for multidimensional networks, heterogeneous nodes (like sellers, products, buyers) with different link labels (like seller-sells-item, buyer-is-friendly, etc.) may exist (Contractor, 2009). Therefore, for multidimensional networks, the link mining must consider the node types with diverse interactions for predicting new links. Also, the CD and social recommendation will be performed differently in multidimensional networks as the interaction between user nodes may differ because of diverse links (Sun & Han, 2012).

Some networks are multidimensional in that they contain nodes of multiple types (e.g., sellers, products, buyers) and links with different labels (e.g., seller, sells-item; buyer, is-friendly; etc.) (Contractor, 2009). The diversity present in multidimensional networks pose special challenges for network analysis, and social recommendation as well as CD for multidimensional networks is an active area of research (Sun & Han, 2012).

In networks where links reflect shared preferences between users (e.g., same purchases, same websites visited, etc.), these shared traits may be attributable to different causes. Social influence and self-selection both play a role in link formation. Self-selection refers to the fact that people with similar traits incline to form social links with each other, while social influence signifies that users, who are linked to each other, tend to adopt similar attributes or preferences (Shalizi & Thomas, 2011).

1.4.4 Research Directions in Ethical Aspects of Data Collection for Business Analytics (Sivarajah et al., 2017)

Business organizations have a serious role to play in safeguarding the data gathered from users to avoid misuse of information. The implications of uncertain ethical conduct when handling big data are vital and can have a catastrophic impact on the reputation and revenues of an organization. The trust of both customers and employees in the business organization can be seriously impaired. In this regard, the following subsections list some ethical issues that need to be handled effectively and which require ongoing research due to the increasing complexity and comprehensiveness of information systems.

1.4.4.1 Safeguarding the Personal Data of Customers

A key ethical issue currently faced by business organizations is related to the conflicting goals of gathering and utilizing information for enhancing the performance and consumer services on the one hand and maintaining the confidentiality of all stakeholders on the other hand. To achieve both goals, several steps may be followed:

- Prevent profiling of customers using high-end technology: Novel tools for surveillance and highly sophisticated information collection methods can gather detailed personal information of customers. This generates an increased threat to the personal privacy of individuals. Hence, due care must be taken by the organizations to utilize these tools in a reliable and accountable fashion.
- Safeguard impartiality in computerized decisions: When decisions are made automatically by computers in sectors such as education, healthcare, etc., chances are high that the decisions made may be unfair to certain weaker section of the society. Therefore, it is the duty of the business organization to perform independent reviews to safeguard the interests of the people who may be economically or socially backward.
- Establish and enforce rules and laws that prevent misuse of collected data. The privacy, freedom, and confidentiality of the customers must be maintained and protected legally.
- Give individuals working in organizations a measure of control over the information collected from them, especially personal data.
- Safeguard individuals from incorrect information: Databases associated with organizations and corporations should permit every person to suitably safeguard the correctness of personal data that will be used to make critical decisions about them. This can be resolved by mandatory disclosure of information to the stakeholders so that incorrect information may be rectified.

1.4.4.2 Customer Profiling

Profiling of customers is usually done to recommend certain products to the consumers based on their choices. This type of data gathered by the organizations is usually done for the purposes of marketing. It can also be utilized to govern the consumer's personal characteristics such as the capability of a person to pay for a particular service or encourage their chances of purchasing real estate properties, etc. However, this type of information can be used in an adverse manner to violate consumers' privacy and expose them to unwanted solicitations. Research is needed to determine effective means for preventing negative impacts and for giving stakeholders a measure of control over the gathering of this type of information.

1.5 Conclusions

In conclusion, ML has a positive impact on the overall business operations and the creation of market leadership. Business organizations that implement ML in their operations can achieve high operation optimization. The adoption of ML improves overall decision-making within an organization by using ML insights to make informed decisions. In marketing, ML is used to ensure an organization marketing effort is not wasted and product information can reach the potential customers. ML has increased automation of business processes and production processes, which reduce overall production costs and help create high-quality products for mass consumption. The use of ML helps business organizations have a proactive approach towards cybersecurity, which improves the security of business and customer information. A detailed discussion of the different techniques of ML and its applications in business analytics is provided in this chapter. The reader of this chapter can be benefitted by the knowledge about the current projects going on in the domain of business analytics by utilizing the techniques of ML. The open research problems along with the challenges faced for applying ML techniques for business analytics is also discussed.

References

6 examples of AI in business intelligence applications. (n.d.). https://emerj.com/ai-sector-overviews/ai-in-business-intelligence-applications/. Accessed online on 2 July 2020.

Abbasi, A., & Chen, H. (2008). CyberGate: A system and design framework for text analysis of computer mediated communication. *MIS Quarterly, 32*(4), 811–837.

Abiodun, O., Jantan, A., Omolara, O., Dada, K., Mohamed, N., & Arshad, H. (2018). State-of-the-art in artificial neural network applications: A survey. *Heliyon, 4*, e00938. https://doi.org/10.1016/j.heliyon.2018.e00938

Agarwal, A., Chakrabarti S., & Aggarwal, S. (2006). Learning to rank networked entities. In *Proceedings of the ACM SIGKDD international conference on knowledge discovery and data mining* (pp. 14–23).

Al Hamad, M., & Zeki, A. M. (2018). Accuracy vs. cost in decision trees: A survey. In *2018 international conference on innovation and intelligence for informatics, computing, and technologies (3ICT)*, Sakhier (pp. 1–4). https://doi.org/10.1109/3ICT.2018.8855780.

Al-Aidaroos, K. M., Bakar, A. A., & Othman, Z. (2010). Naïve bayes variants in classification learning. In *2010 international conference on information retrieval & knowledge management (CAMP)*, Shah Alam, Selangor (pp. 276–281). https://doi.org/10.1109/INFRKM.2010.5466902.

Alam, T. (2019). Forecasting exports and imports through artificial neural network and autoregressive integrated moving average.Decision. *Science Letters, 8*(3), 249–260.

Archana Bai, S. (2011). Machine Learning technologies in business and engineering. In *International conference on sustainable energy and intelligent systems (SEISCON 2011)*, Chennai (pp. 856–859). https://doi.org/10.1049/cp.2011.0486.

Better Bayesian Learning. (2003). http://paulgraham.com/better.html. Accessed 9 Jan 2021.

Bhattacharyya, S., & Bandyopadhyay, G. (2014). Comparative analysis using multinomial logistic regression. In *2014 2nd international conference on business and information management (ICBIM)*, Durgapur (pp. 119–124). https://doi.org/10.1109/ICBIM.2014.6970970.

Brynjolfsson, E., & Mcafee, A. (2017). The business of artificial intelligence. *Harvard Business Review 4* (11), 1–41.

Chen, H., & Zimbra, D. (2010). AI and opinion mining. *IEEE Intelligent Systems, 25*(3), 74–76.

Chua, F. C. T., Lauw, H. W., & Lim E. -P. (2011). Predicting item adoption using social correlation. In *Proceedings of the SIAM international conference on data mining* (pp. 367–378).

Contractor, N. (2009). The emergence of multidimensional networks. *Journal of Computer-Mediated Communication, 14*, 3.

Faggella, D. (2018a, December). *Furniture retailer saves time on customer support with routing and macros*. AI Case Studies.

Faggella, D. (2018b, December). *Smart home services provider automated report creation with AI and customer data*. AI Case Studies.

Faggella, D. (2018c, December). *Microsoft gets the pulse of customer sentiment with natural language processing*. AI Case Studies.

Faggella, D. (2019, January). *Automotive repair equipment OEM uses AI to monetize repair service data*. Business Intelligence and Analytics.

Fortunato, S. (2010). Community detection in graphs. *Physics Reports, 486*(3–5), 75–174.

Gaber, M. M., Zaslavsky, A., & Krishnaswamy, S. (2005). Mining data streams: A review. *SIGMOD Record, 34*(2), 18–26.

Goh, R., & Lee, L. S. (2019, 2019). Credit scoring: A review on support vector machines and metaheuristic approaches. *Advances in Operations Research*, 1–30. https://doi.org/10.1155/2019/1974794

Guha, R., McCool, R., & Miller E. (2003). Semantic search. In *Proceedings of the international world wide web conference*.

Guo-sheng, H., & Guo-hong, Z. (2007). The study of credit evaluation of business websites using support vector machines. In *2007 international conference on management science and engineering*, Harbin (pp. 263–267). https://doi.org/10.1109/ICMSE.2007.4421858.

Gupta, A., Salau, A. O., Chaturvedi, P., Akinola, S. A., & Ikechi Nwulu, N. (2019). Artificial neural networks: Its techniques and applications to forecasting. In *2019 international conference on automation, computational and technology management (ICACTM)*, London (pp. 320–324). https://doi.org/10.1109/ICACTM.2019.8776701.

Haifley, T. (2002). Linear logistic regression: An introduction. In *IEEE international integrated reliability workshop final report*, 2002, Lake Tahoe (pp. 184–187). https://doi.org/10.1109/IRWS.2002.1194264.

Henschen, D. (2011). Why all the Hadoopla? *Information Week, 11*(14/11), 19–26.

Herlocker, J. L., Konstan, J. A., Borchers, A., & Riedl, J. (1999). An algorithmic framework for performing collaborative filtering. In *Proceedings of the 22nd annual international ACM SIGIR conference on research and development in information retrieval* (pp. 230–237).

Hui-lin, Q., & Feng, G. (2011). A research on logistic regression model based corporate credit rating. In *2011 international conference on E-Business and E-Government (ICEE)*, Shanghai (pp. 1–4). https://doi.org/10.1109/ICEBEG.2011.5882285.

IBM. (2011, November15). The 2011 IBM tech trends report. http://ibm.com/developerworks/techntrendsreport. Accessed online on 10 July 2020.

Jinguo, X., & Chen, X. (2011). Application of decision tree method in economic statistical data processing. In *2011 international conference on E-Business and E-Government (ICEE)*, Shanghai (pp. 1–4). https://doi.org/10.1109/ICEBEG.2011.5887040.

Kang, U., Tsourakakis, C. E., Appel, A. P., Faloutsos, C., & Leskovec, J. (2011). HADI: Mining radii of large graphs. *ACM Transactions on Knowledge Discovery from Data, 5*(2), 1–24.

Karamizadeh, S., Abdullah, S. M., Halimi, M., Shayan, J., & Rajabi, M. J. (2014). Advantage and drawback of support vector machine functionality. In *2014 international conference on computer, communications, and control technology (I4CT)*, Langkawi (pp. 63–65). https://doi.org/10.1109/I4CT.2014.6914146.

Kotsiantis, S. B. (2013). Decision trees: A recent overview. *Artificial Intelligence Review, 39*, 261. https://doi.org/10.1007/s10462-011-9272-4

Liben-Nowell, D., & Kleinberg, J. (2007). The link-prediction problem for social networks. *Journal of the American Society for Information Science and Technology, 58*(7), 1019–1031.

Lim, E. P., Chen, H., & Chen, G. (2013). Business intelligence and analytics: Research directions. *ACM Transactions on Management Information Systems, 3*(4), 1–10. Research Collection School of Information Systems.

Lin, J., Snow, R., & Morgan, W. 2011. Smoothing techniques for adaptive online language models: Topic tracking in tweet streams. In *Proceedings of the ACM SIGKDD international conference on knowledge discovery and data mining*.

Lovell, B., & Walder, C. (2006). Support vector machines for business applications. In *Mathematical methods for knowledge discovery and data mining*. https://doi.org/10.4018/978-1-59904-528-3.ch005

Ma, H., Yang, H., Lyu, M. R., & King, I. 2008. Sorec: Social recommendation using probabilistic matrix factorization. In *Proceedings of the ACM conference on information and knowledge management* (pp. 931–940).

Martinez-Arroyo, M., & Sucar, L. E. (2006). Learning an optimal naive bayes classifier. In *18th international conference on pattern recognition (ICPR'06)*, Hong Kong (pp. 1236–1239). https://doi.org/10.1109/ICPR.2006.748.

Mishra, M., & Srivastava, M. (2014). A view of artificial neural network. In *2014 international conference on advances in engineering & technology research (ICAETR - 2014)*, Unnao (pp. 1–3). https://doi.org/10.1109/ICAETR.2014.7012785.

Narula, G. (2019). Machine learning algorithms for business applications – complete guide. Business Intelligence and Analytics.

Page, L., Brin, S., Motwani, R., & Winograd, T. (1999). The PageRank citation ranking: Bringing order to the Web. http://dbpubs.stanford.edu/pub/1999-66.

Parseh, M., Rahmanimanesh, M., & Keshavarzi, P. (2020). Persian handwritten digit recognition using combination of convolutional neural network and support vector machine methods. *The International Arab Journal of Information Technology, 17*, 572–578. https://doi.org/10.34028/iajit/17/4/16

Porter, M. A., Onnela, J.-P., & Mucha, P. J. (2009). Communities in networks. *Notice of AMS, 56*(9), 1082–1097.

Savage, N. (2012). Gaining wisdom from crowds. *Communications of the ACM, 55*(3), 13–15.

Shalizi, C. R., & Thomas, A. C. (2011). Homophily and contagion are generically confounded in observational social network studies. *Sociological Methods & Research, 40*, 211–239.

Sivarajah, U., Kamal, M. M., Irani, Z., & Weerakkody, V. (2017). Critical analysis of big data challenges and analytical methods. *Journal of Business Research, 70*, 263–286.

Sun, Y., & Han, J. (2012). *Mining heterogeneous information networks: Principles and methodologies*. Morgan & Claypool Publishers.

Trippi, R. R., & Turban, E. (1992). *Neural networks in finance and investing: Using Machine Learning to improve real-world performance*. McGraw-Hill, Inc.

Vasundhara, D. N., & Seetha, M. (2016). Rough-set and artificial neural networks-based image classification. In *2016 2nd international conference on contemporary computing and informatics (IC3I)*, Noida (pp. 35–39). https://doi.org/10.1109/IC3I.2016.7917931.

Xie, T., Yu, H., & Wilamowski, B. (2011). Comparison between traditional neural networks and radial basis function networks. In *2011 IEEE international symposium on industrial electronics*, Gdansk (pp. 1194–1199). https://doi.org/10.1109/ISIE.2011.5984328.

Yu, X., Qi, Z., & Zhao, Y. (2013). Support vector regression for newspaper/magazine sales forecasting. *Procedia Computer Science, 17*, 1055–1062. https://doi.org/10.1016/j.procs.2013.05.134

Zahrani, A., & Marghalani, A. (2018). How artificial intelligent transform business? https://doi.org/10.13140/RG.2.2.20426.67522

Zekic-Susac, M., Sarlija, N., & Bensic, M. (2004). Small business credit scoring: A comparison of logistic regression, neural network, and decision tree models. In *26th international conference on information technology interfaces*, 2004, Cavtat (pp. 265–270, Vol. 1).

Zhong, Y. (2016). The analysis of cases based on decision tree. In *2016 7th IEEE international conference on software engineering and service science (ICSESS)*, Beijing (pp. 142–147). https://doi.org/10.1109/ICSESS.2016.7883035.

Chapter 2
Past Achievements and Future Promises of Digital Transformation: A Literature Review

Mohamed Alloghani, Christopher Thron, and Saad Subair

2.1 Introduction

Digital technology has profoundly impacted nearly every aspect of society. In business, it has demonstrated extraordinary capacity for restructuring of operations, processes, and market characteristics and in the process has fundamentally changed industries and disrupting markets. Today, the focus of implementing technology in many settings has moved beyond improvement of efficiency of internal operations to include improvement of interactions with clients and improvement or complete revolution of services. In order to describe these new possibilities, a new terminology has sprung up, including such popular buzz words as 'Internet of Things', 'Industry 4.0', and 'Big Data'. One new term that is becoming increasingly pervasive in industry (and, more recently, among researchers) is 'digital transformation'. As described earlier in the abstract, digital transformation refers to the idea of extensive restructuring of operations in organizations, business agencies, and other entities to incorporate technological innovations that completely reshape the approach to the processes. As depicted from Fig. 2.1, the digital transformation integrates

M. Alloghani (✉)
Computing and Mathematical Sciences, Liverpool John Moores University, Liverpool, UK

The UAE Artificial Intelligence Office, Prime Minister's Office at the Ministry of Cabinet Affairs and the Future, Dubai, UAE

C. Thron
Department of Science and Mathematics, Texas A&M University-Central Texas, Killeen, TX, USA
e-mail: thron@tamuct.edu

S. Subair
College of Computer Studies, International University of Africa, Khartoum, Sudan
e-mail: ssubair@iua.edu.sd

M. Alloghani et al. (eds.), *Artificial Intelligence for Data Science in Theory and Practice*, Studies in Computational Intelligence 1006,
https://doi.org/10.1007/978-3-030-92245-0_2

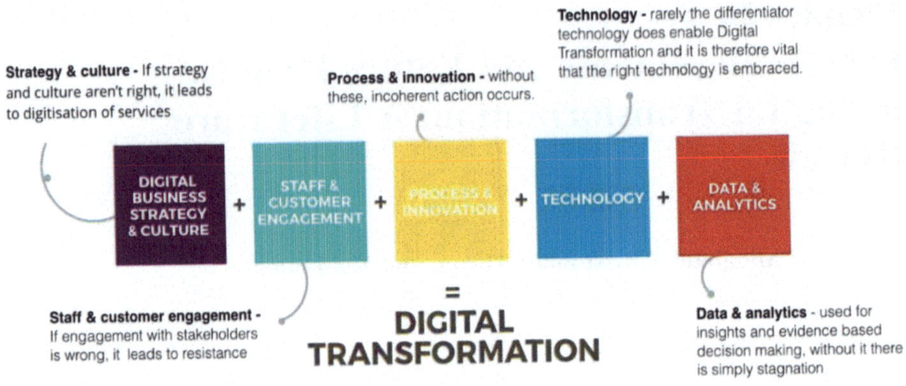

Fig. 2.1 Key blocks of digital transformation (Ionology, 2020)

digital technology into all aspects of an entity's operations and delivering value to customers. It's also a cultural shift that necessitates organizations challenging the status quo, experimenting, and becoming agile. The design, business model, and operations are all included in the three-stage process of digital transformation.

This concept is still relatively new, and the literature on its description, adoption, and impact is lacking in many aspects. This literature review aims to address these shortcomings by identifying scholarly research and information that currently exists regarding the conceptualization, adoption, impacts, and future trends of digital transformation.

2.2 Materials and Methods

To map out as much of the existing information on the topic as possible, a systematic review of the literature was conducted. This approach involved a strategic search and identification of resources that addressed various parts of the topic, followed by a careful review of the contents of the published resources and the ideas or arguments presented in each. The data collection section of a systematic review typically entails deliberate steps and procedures that are described and performed with sufficient clarity as to allow their replication. As a guide for searching for resources for this review, the author applied selected PRISMA, a framework developed to guide the systematic identification and acceptance or rejection of scholarly articles from databases and other sources.

2.2.1 Search Strategy

The author conducted a search for literature on the topic in two of the largest databases of peer-reviewed articles, abstracts, and other resources in several fields: Web of Science and Scopus. The search strategy used combinations of carefully selected keywords to retrieve published works related to the topic of interest. The specific keywords utilized were: 'digital transformation', 'IT-driven transformation', 'IT-enabled business transformation', 'digitization', and 'digital business strategy'. To increase chances of retrieving a larger number of relevant publications, various combinations of these search terms were applied alternatively using Boolean operators 'AND' and 'OR'. Since the literature review was intended to capture as much of the relevant information that already exists on the topic as possible, there was no specification on the date of publication of the resources to be retrieved. However, the type of sources was specified as journal articles and conference proceedings only, a limitation that was applied to exclude sources that may be too lengthy to allow sufficient review or contain non-scholarly information and poorly researched or non-peer-reviewed data. The search terms and specifications used were similar for both databases. Although the database search constituted the primary source of articles for the review, the author decided to review the bibliography sections of articles that meet inclusion criteria for possibly relevant sources. After running the keywords in isolation and as Boolean combinations and returning results, the author performed a manual scan of all the titles and abstracts to determine relevance of each source, which was dictated by the inclusion and exclusion criteria.

2.2.2 Inclusion and Exclusion Criteria

Although the author was cautious about excluding useful publications or ending up with too few resources for the review, certain basic criteria had to be met for a source to be included. To avoid the potential for misinformation and inaccurate reporting, all sources that would be included had to be written in English language. Additionally, the abstract, titles, or keywords of the publications that would be selected should mention at least one of the keywords used to obtain the sources from the databases. However, some resources only used these keywords as entry terms in the introductory or abstract section before proceeding to discuss other unrelated issues. Such articles were rejected.

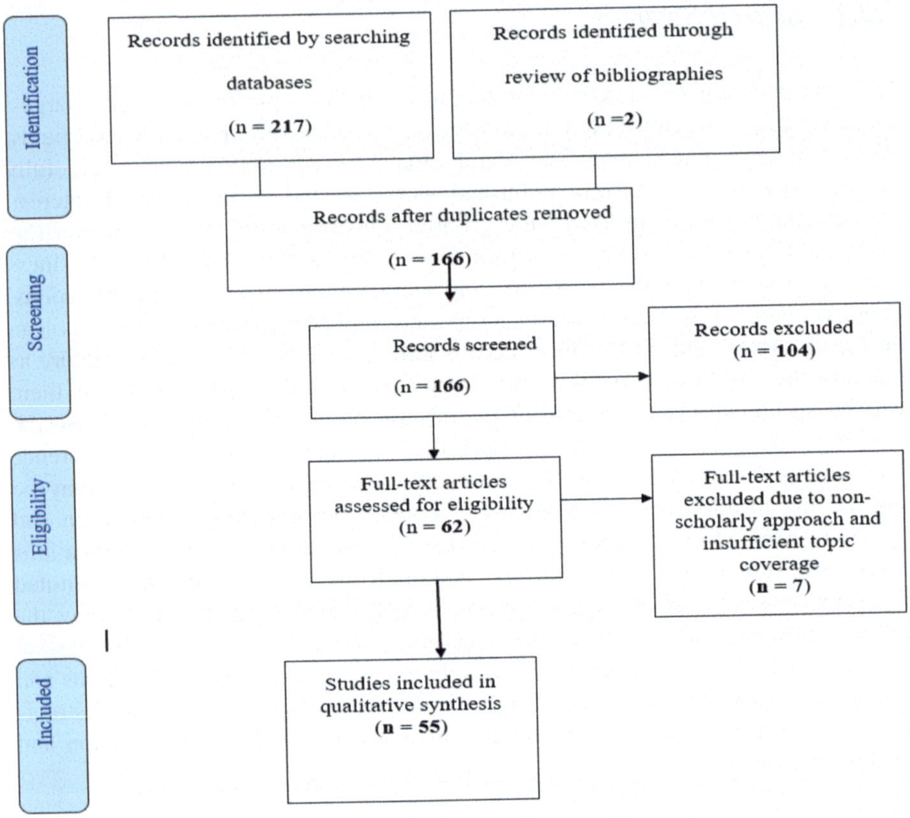

Fig. 2.2 PRISMA flow diagram showing number of records identified, screened, and included in the literature review

2.3 Results and Discussion

As shown in Fig. 2.2, the initial search from the two databases returned a combined 217 articles from different journals and conference proceedings. Comparison of the two results lists revealed that 53 articles were duplicates and their exclusion left 164 articles. The screening of the titles and abstracts of the remaining articles resulted in the exclusion of 104 articles, most of which were written in French and German. Some 60 articles were then read in their entirety to determine their appropriateness for the review. Three of these sources presented discussions in formats and tones that were not consistent with scholarly research. A further four articles covered too little of digital transformation as the subject to enable meaningful contribution to the topic. Two relevant articles were found by analysing the bibliographies of the included sources. Eventually, 55 articles were included for the qualitative and quantitative analysis that formed part of the current review. The PRISMA flow diagram for the search process and results is shown in Fig. 2.2.

The sources selected for the final review discussed a large amount of information on a wide range of topics related to digital transformation. Additionally, the articles displayed a number of interesting characteristics that pointed to the nature of the topic under research. For instance, despite the lack of restriction on the date of publication during the search for information, the oldest resource included for the review was written in 2010 (Agarwal et al., 2010). The article, published in the Information Systems Research journal, presents one of the earliest reviews of existing literature on digital transformation. However, like many of the previous reviews of literature on the issue, the article's scope is restricted to a specific component of digital transformation—in this case, its adoption in the health sector of the United States (Agarwal et al., 2010). Another two articles were published 5 years later, both discussing the theoretical approaches to adoption of digital strategies in organization (Ganguly, 2015; Schuchman & Seufert, 2015). Remarkably, the bulk of the resources utilized for this review were published in 2018 (20, 36.4%) and 2020 (18, 32.7%), up from only nine sources (16.4%) published in 2017 (Table 2.1). This sharp increase in the number of academic articles released in the last 2 years suggests the emergence of important factors driving scholarly interest in the phenomenon of digital transformation.

The search for distinct drivers of such an influx of academic publications on the matter did not reveal specific causes, although a significant portion of the sources published in the 2 years were presented in conferences held in different countries in Europe during the same period.

Another interesting pattern that is apparent in the existing body of literature on digital transformation is the concentration of published articles in the European continent. As part of the quantitative analysis, the sources were categorized according to the country or region on which the discussion concentrates, which was often the same country where the article was published. While many of the sources clearly identified the scope of their research or discussion as a specific country, some addressed the issue of digital transformation from a general viewpoint, without making specific references or examples of local situations in any country. Still, others discussed the topic from the perspective of an entire continent rather than specific countries. Despite these challenges, the analysis enabled the identification of countries and regions with the highest number of publications on the topic. Fourteen of the sources (25.5%) were either published in or discussed the situation in Germany (Table 2.2). Most of the other articles were either associated with another country in Europe or addressed aspects of digital transformation in the European continent in general.

This concentration of published literature in the European continent has not been addressed in previous literature reviews. Although an explanation for the skew has not been formally sought, the difference in the distribution of resources on the topic could be explained in two ways. The first explanation could be that the keywords and search terms used corresponded to terminology most widely used in European countries. This situation is highly likely since 'digital transformation' and 'digitization' are terms coined in informal conversations and literature in many European countries, while in the United States and other countries in the American

Table 2.1 Distribution of sources over time

Year	Number of sources	Percentage	References
2010	1	1.8	Agarwal et al. (2010)
2015	2	3.6	Ganguly (2015), Schuchman and Seufert (2015)
2016	5	9	Stief et al. (2016), Hess et al. (2016), Majchrzak et al. (2016), Henriette et al. (2016), OECD (2016)
2017	9	16.4	Reddy and Reinartz (2017), Paritala et al. (2017), von Leipzig et al. (2017), Omar et al. (2017), Parviainen et al. (2017), Ndemo and Weiss (2017), Schallmo et al. (2017), Resego et al. (2017), European Union (2017)
2018	20	36.4	Al-Ruithe et al. (2018), Goerzig and Bauernhansl (2018), Zahara and Petreanu (2018), Issa et al. (2018), Osmundsen et al. (2018), Afonasova (2018), Rachinger et al. (2018), Hess and Constantiou (2018), Ibarra et al. (2018), Kotarba (2018), Skog et al. (2018), Limani et al. (2018), Benjamin and Potts (2018), Nadeem et al. (2018), Sow (2018), Chanias et al. (2018), Nwaiwu (2018), Kirsten et al. (2018), Tewes et al. (2018), Moreira et al. (2018)
2020	18	32.7	Oertwig et al. (2019), Genzorova et al. (2019), Mergel et al. (2019), Vogelsang et al. (2019), Dugstad et al. (2019), Zulkarnain et al. (2019), Mhlungu et al. (2019), Gbadegeshin (2019), Savastano et al. (2019), Dufva and Dufva (2019), Cortellazzo et al. (2019), Holth and Boe (2019), Promsri (2019), Kaplan and Tewes (2019), Verina and Titko (2019), Ivancic et al. (2019), Pelletier and Cloutier (2019), Skog (2019)

continent, the term ‘Internet industry’ has often been applied as an equivalent to ‘Industry 4.0’, another connotation of the widespread infiltration of technological alterations of operations across industries (Dufva & Dufva, 2019). Therefore, the large number of studies in European continent could reflect a bias in the search strategy towards this region. Alternatively, the skewness is a representation of the industries and sectors most affected by digital transformation. For instance, the sources from European countries addressed the incorporation of digital innovations in manufacturing, engineering, banking and commerce, retail, government services, and other sectors, while US-based articles were largely restricted to the adoption of digital technologies in the health sector and the role of leadership in the incorporation of technological strategies. The review of the selected scholarly resources revealed that digital transformation is a wide concept with several components and dimensions.

The bulk of studies now seem interested in the dynamics of adoption of digital transformation in organizations, a tendency that reflects the interests of practitioners

Table 2.2 Classification of sources by country/region

Country/region	No. of sources	Percentage	References
Germany	14	25.5	Oertwig et al. (2019), Goerzig and Bauernhansl (2018), von Leipzig et al. (2017), Vogelsang et al. (2019), Stief et al. (2016), Reddy and Reinartz (2017), Gbadegeshin (2019), Hess and Constantiou (2018), Savastano et al. (2019), Schallmo et al. (2017), Kaplan and Tewes (2019), Chanias et al. (2018), Kirsten et al. (2018), Tewes et al. (2018)
United Kingdom	4	7.3	Omar et al. (2017), Issa et al. (2018), Ganguly (2015), Benjamin and Potts (2018)
United States	7	12.7	Dugstad et al. (2019), Agarwal et al. (2010), Zulkarnain et al. (2019), Hess et al. (2016), Rachinger et al. (2018), Majchrzak et al. (2016), Sow (2018)
Canada	2	3.6	Ivancic et al. (2019), Pelletier and Cloutier (2019)
Europe	3	5.4	Mergel et al. (2019), OECD (2016), European Union (2017)
Romania	2	3.6	Zahara and Petreanu (2018), Ibarra et al. (2018)
Finland	2	3.6	Parviainen et al. (2017), Dufva and Dufva (2019)
Switzerland	1	1.8	Schuchman and Seufert (2015)
France	1	1.8	Henriette et al. (2016)
Greece	1	1.8	Osmundsen et al. (2018)
Portugal	1	1.8	Moreira et al. (2018)
Russia	1	1.8	Afonasova (2018)
South Africa	1	1.8	Mhlungu et al. (2019)
Kenya	1	1.8	Ndemo and Weiss (2017)
Poland	1	1.8	Kotarba (2018)
Norway	1	1.8	Holth and Boe (2019)
Italy	1	1.8	Cortellazzo et al. (2019)
Sweden	2	3.6	Skog et al. (2018), Skog (2019)
Australia	2	3.6	Paritala et al. (2017), Nadeem et al. (2018)
Kosovo	1	1.8	Limani et al. (2018)
Thailand	1	1.8	Promsri (2019)
Lithuania	1	1.8	Verina and Titko (2019)
Czech Republic	1	1.8	Nwaiwu (2018)
Slovenia	1	1.8	Resego et al. (2017)
Saudi Arabia	1	1.8	Al-Ruithe et al. (2018)
Slovak Republic	1	1.8	Genzorova et al. (2019)

Table 2.3 Classification of sources by research areas

Research area	Number of sources	Percentage	References
Definition and conceptualization	8	14.5	Mergel et al. (2019), Hess et al. (2016), Skog et al. (2018), Schallmo et al. (2017), Verina and Titko (2019), Nwaiwu (2018), Resego et al. (2017), European Union (2017)
Adoption of digital transformation	34	61.8	Oertwig et al. (2019), Al-Ruithe et al. (2018), Genzorova et al. (2019), Goerzig and Bauernhansl (2018), von Leipzig et al. (2017), Zahara and Petreanu (2018), Omar et al. (2017), Issa et al. (2018), Osmundsen et al. (2018), Vogelsang et al. (2019), Dugstad et al. (2019), Agarwal et al. (2010), Stief et al. (2016), Zulkarnain et al. (2019), Afonasova (2018), Mhlungu et al. (2019), Hess and Constantiou (2018), Parviainen et al. (2017), Ndemo and Weiss (2017), Majchrzak et al. (2016), Holth and Boe (2019), Limani et al. (2018), Henriette et al. (2016), Benjamin and Potts (2018), Nadeem et al. (2018), Sow (2018), Promsri (2019), Schuchman and Seufert (2015), Ivancic et al. (2019), Chanias et al. (2018), Kirsten et al. (2018), Pelletier and Cloutier (2019), Skog (2019), OECD (2016)
Impact of digital transformation	8	14.5	Rachinger et al. (2018), Reddy and Reinartz (2017), Gbadegeshin (2019), Ibarra et al. (2018), Savastano et al. (2019), Kotarba (2018), Cortellazzo et al. (2019), Moreira et al. (2018)
Future trends in digital transformation	5	9.1	Paritala et al. (2017), Dufva and Dufva (2019), Ganguly (2015), Kaplan and Tewes (2019), Tewes et al. (2018)

whose focus is on implementing digital solutions rather than studying what they mean. Altogether, the scholarly literature may be categorized into four general areas: definition and conceptualization of digital transformation (14.5% of sources), adoption of digital solutions (including incorporation strategies, challenges, and success factors) (61.8% of sources), impact of digital transformation (on different areas of organizational operation such as business models) (14.5% of sources), and future trends in digital transformation (9.1% of sources). Table 2.3 shows the distribution of articles between these research areas.

One of the recurrent themes in the research on the adoption of digital strategies is the need to view this phenomenon as a full-scale organizational change, as opposed to treating it as a technological innovation that needs to be incorporated into the company's procedures. Therefore, researchers recommend the mobilization of the organization's entire resources, especially the extensive involvement of human resources in digitization efforts. Other variables identified as critical determinants of success in transformation operations include leadership, collaboration with suppliers and customers, adoption of an agile approach, and organization culture change.

The discussions on adoption of digital processes are dominated by theoretical frameworks that provide guidelines on appropriate digitization strategies. Articles that fall into the 'adoption of digital transformation' category also outline important challenges to the effective incorporation of technological innovations in organizational operations, with the most commonly cited barriers including cultural inertia, lack of strategic management, inadequate human resource involvement, and lack of defined digital strategies (Zahara & Petreanu, 2018; Afonasova, 2018; Parviainen et al., 2017). Among these theoretical models, the most emphasized approaches are resource-based—which encourage organizations to accumulate and mobilize resources towards the digital change and value proposition—which emphasize the need to utilize digital processes to generate value for the client. The latter theoretical model is particularly common among organizations in the manufacturing sector, where researchers have repeatedly acknowledged the preference by clients of customization of products as the most important measures of value. Many studies that addressed the impact of digital transformation on the business models of the companies that are affected found that value proposition is the most convenient approach to utilize the change for income generation (Genzorova et al., 2019; Rachinger et al., 2018; Ibarra et al., 2018; Tewes et al., 2018).

The most recent of these drivers include Internet of Things, artificial intelligence (AI), 3D printing, Big Data, drones and robotics, and augmented and virtual reality (Paritala et al., 2017; Dufva & Dufva, 2019; Ivancic et al., 2019; European Union, 2017). Artificial intelligence (AI) is considered one of the most disruptive emerging technologies, although its implementation and development is still at its infant stages (Dufva & Dufva, 2019; Verina & Titko, 2019). Presently, the most significant impact of AI is in machine learning, where application of statistical algorithms and patterns allows users to automate a remarkably large number of tasks and create self-organizing systems (Dufva & Dufva, 2019; Schallmo et al., 2017). This functionality is particularly useful in manufacturing industries, where automation of repetitive tasks has massive effects on the cost of production and value creation. The automation afforded by classical AI technologies also reduces the rates of errors in processes and increases operation speed.

Articles that focus on the definition and conceptualization of digital transformation uniformly point out that this aspect of the phenomenon has been largely neglected. Despite the widespread discussion and publication of items on the adoption strategies, driving forces, success factors, and challenges facing the incorporation of digital processes in organizations, little is still known about what digital transformation itself means. A commonly expressed concern is that there are too many definitions of digital transformation, many of which are vague and do not reflect the full extent of components encompassed by the process. Conceptual frameworks that attempt to describe what digital transformation is generally break it down into component that correspond to the digital technologies being adopted and the user experience (Henriette et al., 2016). An important third dimension is organizational human resources, which is increasingly being emphasized as a critical determinant of success in the transformation process.

Discussions of the future trends in digital transformation are not abundant. Among the articles that address this question, the focus is on appropriate adoption strategies for the upcoming technologies, as well as the expected impact of the changes on various areas of organizational operations. Researchers tend to agree that the future will bring along increased adoption of technology in enterprises that already embrace innovations and a forceful permeation into firms that currently resist the change, which will be forced to adopt the strategy in order to survive in the fiercely competitive business environment that will result.

2.4 Conclusions

The widespread influence of technological innovations in business, education, public service, and other areas of society has sparked a large amount of interest among practitioners in these sectors. Additionally, this rising significance of technology, particularly its capacity to radically change approaches to service delivery, organizational operations, and other functional aspects of businesses means that digital transformation has the potential to reshape the way business is conducted, which has implications for both enterprises and their clients. Consequently, there have been increased attempts to understand various theoretical concepts related to the widespread adoption of technology in operations, including digital transformation, digitization, Internet of Things, and Big Data. Although a considerable amount of literature already exists on digital transformation, a systematic review of this content reveals that most of it has been published in the last 2 years. This trend suggests that the topic is still new and has recently generated a high amount of both academic and practitioner interest. Despite the now large amount of literature on digital transformation, the term still remains to be adequately conceptualized, and definitions of the concept are many and often vague. Future research in this area should focus on consolidating the conceptualization efforts to develop a clear definition and a single, comprehensive theoretical framework for approaching digital framework.

References

Afonasova, M. (2018). Digital transformation of the entrepreneurship: Challenges and prospects. *Journal of Entrepreneurship Education, 21*(25), 1–7.

Agarwal, R., Gao, G., DesRoches, C., & Jha, A. (2010). The digital transformation of healthcare: Current status and the road ahead. *Information Systems Research, 21*(4), 796–809.

Al-Ruithe, M., Benkhelifa, E., & Hameed, K. (2018). Key issues for embracing the cloud computing to adopt a digital transformation: A study of Saudi public sector. *Procedia Computer Science, 130*, 1037–1043.

Benjamin, K., & Potts, W. (2018). Digital transformation in government: Lessons for digital health? *Digital Health, 3*(1), 1–5.

Chanias, S., Myers, M., & Hess, T. (2018). Digital transformation strategy making in pre-digital organizations: The case of a financial services provider. *Journal of Strategic Information Systems*, 1–17. https://doi.org/10.1016/j/jsis.2018.11.003

Cortellazzo, L., Bruni, E., & Zampieri, R. (2019). The role of leadership in a digitized world: A review. *Frontiers in Psychology, 10*(1938), 2–21.

Dufva, T., & Dufva, M. (2019). Grasping the future of the digital society. *Futures, 107*(2019), 17–28.

Dugstad, J., Eide, T., Nilsen, E. R., & Eide, H. (2019). Towards successful digital transformation through co-creation: A longitudinal study of a four-year implementation of digital monitoring technology in residential care for persons with dementia. *BMC Health Services Research, 19*(2019), 1–17.

European Union. (2017). *A concept paper on digitization, employability, and inclusiveness: The role of Europe*. DG Communications Networks, Content & Technology.

Ganguly, A. (2015). Optimization of IT and digital transformation: Strategic imperative for creating a new value delivery mechanism and a sustainable future in organization! *European Journal of Business and Innovation Research, 3*(2), 1–13.

Gbadegeshin, S. (2019). The effect of digitalization on the commercialization process of high-technology companies in the life sciences industry. *Technology Innovation Management Review, 9*(1), 49–65.

Genzorova, T., Corejova, T., & Stalmasekova, N. (2019). How digital transformation can influence business model, Case study for transport industry. *Transportation Research Procedia, 40*(2019), 1053–1058.

Goerzig, D., & Bauernhansl, T. (2018). Enterprise architectures for the digital transformation in small and medium-sized enterprises. *Procedia CIRP, 67*(2018), 540–545.

Henriette, E., Feki, M., & Boughzala, I. (2016). Digital transformation challenges. In *Mediterranean Conference on Information Sciences 2016 Proceedings*, Association for Information Systems.

Hess, T., & Constantiou, I. (2018). Introduction to the special issue on "Digitalization and the Media Industry". *Electronic Markets, 28*(2018), 77–78.

Hess, T., Matt, C., Benlian, A., & Wiesbock, F. (2016). Options for formulating a digital transformation strategy. *MIS Quartely Executive, 15*(2), 123–140.

Holth, T., & Boe, O. (2019). Lost in Transition: The dissemination of digitization and the challenges of leading in the military educational organization. *Frontiers in Psychology, 10*(2049), 1–25.

Ibarra, D., Ganzarain, J., & Igartua, J. (2018). Business model innovation through industry 4.0: A review. *Procedia Manufacturing, 22*(2018), 4–10.

Ionology. (2020). Digital transformation framework. [Online]. Available: https://www.dxlatest.com/

Issa, A., Hatiboglu, B., Bildstein, A., & Bauernshansl, T. (2018). Industrie 4.0 roadmap: Framework for digital transformation based on the concepts of capability maturity and alignment. *Procedia CIRP, 72*(2018), 973–978.

Ivancic, L., Vuksic, V., & Spremic, M. (2019). Mastering the digital transformation process: Business practices and lessons learned. *Technology Innovation Management Review, 9*(2), 36–52.

Kaplan, C., & Tewes, S. (2019). Redesigning business model strategy: The digital future of retailing in Europe. *Journal of International Business Research and Marketing, 4*(3), 1–7.

Kirsten, L., Vogelsang, K., Packmohr, S., & Hoppe, U. (2018). Towards a framework for digital transformation success in manufacturing. In *Proceedings of the Twenty-Sixth European Conference on Information Systems (ECIS2018)*, Portsmouth.

Kotarba, M. (2018). Digital transformation of business models. *Foundations of Management, 10*(2018), 123–143.
Limani, Y., Stapleton, L., & Groumpos, P. (2018). The challenges of digital transformation in post-conflict transition regions: Digital technology adoption in Kosovo. *International Federation of Automatic Control, 51-30*, 186–191.
Majchrzak, A., Markus, M., & Wareham, J. (2016). Designing for digital transformation: Lessons for information systems research from the study of ICT and societal challenge. *MIS Quarterly, 40*(2), 267–277.
Mergel, I., Edelmann, N., & Haug, N. (2019). Defining digital transformation: Results from expert interviews. *Government Information Quarterly*, 1–6. https://doi.org/10.1016/j.giq.2019.06.002
Mhlungu, N., Chen, J., & Alkema, P. (2019). The underlying factors of a successful organizational digital transformation. *South African Journal of Information Management, 21*(1), 1–10.
Moreira, F., Ferreira, M., & Seruca, I. (2018). Enterprise 4.0 – the emerging digital transformed enterprise? *Procedia Computer Science, 138*(2018), 525–532.
Nadeem, A., Abedin, B., Cerpa, N., & Chew, E. (2018). Digital transformation & digital business strategy in electronic commerce - the role of organizational capabilities. *Journal of Theoretical and Applied Electronic Commerce Research, 13*(2), 1–8.
Ndemo, B., & Weiss, T. (2017). Making sense of Africa's emerging digital transformation and its many futures. *Africa Journal of Management, 3*(4), 328–347.
Nwaiwu, F. (2018). Review and comparison of conceptual frameworks on digital business transformation. *Journal of Competitiveness, 10*(3), 86–100.
OECD. (2016). *Digital government strategies for transforming public services in the welfare areas*. OECD Comparative Study, OECD.
Oertwig, N., Gering, P., Knothe, T., & Rimmelspacher, S. O. (2019). User-centric process management system for digital transformation of production. *Procedia Manufacturing, 33*(2019), 446–453.
Omar, A., Weerakkody, V., & Sivaraja, U. (2017). Digitally enabled service transformation in UK public sector: A case analysis of universal credit. *International Journal of Information Management, 37*(4), 350–356.
Osmundsen, K., Iden, J., & Bygstad, B. (2018). Digital transformation: Drivers, success factors, and implications. In *Proceedings of the 12th Mediterranean Conference on Information Systems (MCIS)*, Corfu.
Paritala, P., Manchikatla, S., & Yarlagada, P. (2017). Digital manufacturing- applications past, current, and future trends. *Procedia Engineering, 174*, 982–991.
Parviainen, P., Kaariainen, J., Tihinen, M., & Teppola, S. (2017). Tackling the digitalization challenge: How to benefit from digitalization in practice. *International Journal of Information Systems and Project Management, 5*(1), 63–77.
Pelletier, C., & Cloutier, L. (2019). Challenges of digital transformation in SMEs: Exploration of IT-related perceptions in a service ecosystem. In *Proceedings of the 52nd Hawaii International Conference on System Sciences*, Hawaii.
Promsri, C. (2019). Developing model of digital leadership for a successful digital transformation. *International Journal of Business Management, 2*(8), 1–8.
Rachinger, M., Rauter, R., Muller, C., Vorraber, W., & Schirgi, E. (2018). Digitalization and its influence on business model innovation. *Journal of Manufacturing Technology Management*, 1–18. https://doi.org/10.1108/JMTM-01-2018-0020
Reddy, S., & Reinartz, W. (2017). Digital transformation and value creation: Sea change ahead. *Value in the Digital Era, 9*(1), 10.
Resego, M., Audrey, G., & Philip, O. (2017). Conceptualizing digital transformation in business organizations: A systematic review of literature. In *Proceedings of the 30th Bled eConference*, Slovenia, Bled.
Savastano, M., Amendola, C., Bellini, F., & D'Ascenzo, F. (2019). Contextual impacts on industrial processes brought by the digital transformation of manufacturing: A systematic review. *Sustainability, 11*(891), 1–38.

Schallmo, D., Williams, C., & Boardman, L. (2017). Digital transformation of business models — best practice, enablers, and roadmap. *International Journal of Innovation Management, 21*(8), 1–17.

Schuchman, D., & Seufert, S. (2015). Corporate learning in times of digital transformation: A conceptual framework and service portfolio for the learning function in banking organizations. *iJAC, 8*(1), 31–40.

Skog, D. (2019). *The dynamics of digital transformation the role of digital innovation, ecosystems and logics in fundamental organizational change*. Umea University.

Skog, D., Wimelius, H., & Sandberg, J. (2018). Digital disruption. *Business Information Systems Engineering, 605*(5), 431–437.

Sow, D. (2018). Impact of leadership on digital transformation. *Business and Economic Research, 8*(3), 140–146.

Stief, S., Theresa, A., & Voeth, M. (2016). Transform to succeed: An empirical analysis of digital transformation in firms. *World Academy of Science Engineering and Technology, 10*(6), 1833–1844.

Tewes, S., Tewes, C., & Jager, C. (2018). The 9×9 of future business models. *International Journal of Innovation and Economic Development, 4*(5), 39–48.

Verina, N., & Titko, J. (2019). Digital transformation: Conceptual framework. In *Proceedings of the International Scientific Conference on Contemporary Issues In Business, Management And Economics Engineering, Vilnius, Lithuania*. VGTU Press.

Vogelsang, K., Liere-Netheler, K., Packmohr, S., & Hoppe, U. (2019). Success factors for fostering a digital transformation in manufacturing companies. *Journal of Enterprise Transformation*, 1–22. https://doi.org/10.1080/19488289.2019.1578839

von Leipzig, T., Gamp, M., Manz, D., et al. (2017). Initialising customer-orientated digital transformation in enterprises. *Procedia Manufacturing, 8*(2017), 517–524.

Zahara, S. E., & Petreanu, C. V. (2018). Challenges in airport digital transformation. *Transportation Research Procedia, 35*(2018), 90–99.

Zulkarnain, N., Hidayanto, A., & Prabowo, H. (2019). The critical success factors for big data adoption in government. *International Journal of Mechanical Engineering and Technology, 10*(3), 864–875.

Chapter 3
Algorithms for the Development of Deep Learning Models for Classification and Prediction of Learner Behaviour in MOOCs

José Edmond Meku Fotso, Bernabe Batchakui, Roger Nkambou, and George Okereke

3.1 Introduction

3.1.1 Background

The demand for training has grown rapidly in recent years. This is evidenced by the high number of learners at every level and type of education: not only in universities but also at the primary and secondary levels as well as in technical and vocational education, both formal and informal. In order to address the issue of massification in education and to encourage sharing of knowledge, distance learning, and especially e-learning, appears to be a suitable approach. This trend is supported by the international agenda, including UNESCO's "Education for All" initiative, the United Nations' fourth sustainable development goal (SDG4) (UNESCO et al., 2016; UNESCO, 2005), the Incheon declaration on SDG4-Education 2030, and the Continental Education Strategy for Africa (CESA 16-25) (AFRICAINE, 2016). In particular, SDG4's objective is to provide "inclusive and equitable quality education and promote lifelong learning opportunities for all" (UNESCO et al., 2016). Massive open online courses (MOOCs) appear to be a suitable approach to support such an initiative. However, MOOCs face the unsolved

J. E. M. Fotso (✉) · B. Batchakui
National Advanced School of Engineering of Yaounde, University of Yaounde 1, Yaounde, Cameroon

R. Nkambou
Computer Science Department, University of Quebec Montreal, Montreal, QC, Canada

G. Okereke
Computer Science Department, University of Nigeria Nsukka, Nsukka, Nigeria
e-mail: george.okereke@unn.edu.ng

M. Alloghani et al. (eds.), *Artificial Intelligence for Data Science in Theory and Practice*, Studies in Computational Intelligence 1006,
https://doi.org/10.1007/978-3-030-92245-0_3

major problems of high dropout, low completion, and a low success rate. Around 90% of students who enrol in a MOOC fail to complete it (Andres et al., 2018). In addition, while slightly more than half of students intend to receive a certificate of completion from a typical MOOC, only around 30% of these respondents achieve this certification (Brooks et al., 2015).

Much research has addressed the problems of dropout and failure in MOOCs. Because of the high number of learners and their heterogeneity, a huge volume of data is generated by learners' activities. Many models have been designed to predict dropout, completion, certification, and/or success, with dropout prediction being the most common (UNESCO, 2005). However, the concept of dropping out and success needs to be reconsidered in the context of MOOCs because not all the learners enrolled in MOOCs intend to get the certificate or even complete the course. Therefore, for people who enrol in MOOCs for other purposes, than to get the certificate or to complete the course, not completing the course, and/or not getting the certificate should not be considered a failure and thus should not be classified as such. Nevertheless, whatever the learners' objectives, they need to take part in the learning process in order to truly fulfil those objectives.

3.1.2 Interest

Our work aims to predict learner participation in the course learning process. To do this, our objective is to design a model to classify and predict learner behaviour, more specifically learners' interaction in the learning process, including with course activities and resources. Such classification and prediction of learner behaviour can serve many purposes. It can be used to improve personalized support and interventions by course instructors and managers; it can also guide the development of adaptive content and learner pathways for learners (Gardner & Brooks, 2018). Altogether, it can then be used to help predict and prevent dropout, thus improving the completion and success rate.

3.1.3 State of the Art

Prior academic and commercial studies of MOOCs have established that there is a strong correlation between student dropout, student general learning outcomes, and student's behaviour vis-a-vis course activities such as attempting quizzes, posting in forums, submitting homework, and utilizing course resources such as videos, audio lectures, and downloadable files (Brown et al., 2015). These behaviours can be divided into two main types: pure learning behaviour, which involve student–system interactions (e.g., completing quizzes, watching videos) (Sinha et al., 2014), and social behaviours, which involve student–student interactions (e.g., posting or commenting on messages in a forum) (Pursel et al., 2016).

Many research works on learner performance prediction using activity logs have demonstrated that learner's behaviour during the learning process can efficiently serve as learner performance predictor (Brown et al., 2015). Activity logs and social metrics include various aspects of learner behaviour. Therefore, by combining the features extracted from those two main sources with other features, like features from geographical, academic, and socio-professional background, we may be able to obtain a very comprehensive understanding of the learner behaviour and then improve our ability to predict learner behaviour or learner participation in the learning process.

3.1.4 Our Contribution

This chapter describes the development and implementation of more accurate behaviour prediction models for learners enrolled in MOOCs, which are based on deep learning algorithms. Since time series data is involved, recurrent neural networks (RNNs) are used. We compare three RNN architectures: simple RNNs, gated recurrent unit RNNs (GRU RNNs), and long short-term memory (LSTM). We find that simple RNNs provide best prediction performance. Finally, we propose a tool to support efficiently the course designers in their process of supporting and guiding the learners in the learning process.

3.1.5 Structure of the Document

This document is organized as follows. Related prior research is described in Sect. 3.2. The research methodology, including the approach, context, and method, is presented in Sect. 3.3, as well as presentation and analysis of results. Section 3.4 gives the conclusion and the future work or perspectives.

3.2 Related Work

3.2.1 Outline

This section aims to explore the concept, concepts, as well as the state of the art related to prediction of student success, and for other learning outcomes. The learning analytics (LA) techniques are designed to analyse learning-driven data. LA includes (1) descriptive analysis (what happened?), (2) predictive analysis (what will happen next?), (3) diagnostic analysis (why did it happen?), and (4) prescriptive analysis (what should be done to improve?). In our current study related to learner behaviour in MOOCs, the focus is done on the predictive analysis. The section starts with an introduction to learning analytics, and then address the importance of student success predictive models in MOOCs, after that we explore the diverse types of inputs that are used by student success predictive models as well as the various data sources providing those inputs/features. The section also explores the features engineering, helping to extract features from data sources, and the section presents the relation between types of model and the outcome predicted, as well as the algorithms used for predictive models and metrics for their evaluation. The section ends with lessons learned from this exploratory exercise.

3.2.2 Overview of Learning Analytics

The subject matter of this chapter falls under the area of learning analytics (LA) in e-learning on MOOC platforms. LA includes predictive, diagnostic, and prescriptive analysis in addition to descriptive analysis. According to the Society for Learning Analytics Research, "LA is the measurement, collection, analysis and reporting of data about learners and their contexts, for purposes of understanding and optimizing learning and the environments in which it occurs" (Siemens & Baker, 2012). LA utilizes raw data extracted from any learning system, including learning management systems (LMSs), open educational resources, online libraries, e-portfolio systems, and student services systems. The analysis of this data highlights relation between variables in log files that are related to the learning process and generates new knowledge about students' behaviour.

The field of LA is itself a subset of educational data mining (EDM) and consists of four main categories of analysis: (1) descriptive (what happened?), (2) predictive (what will happen next?), (3) diagnostic (why did it happen?), and (4) prescriptive (what should be done to improve?) (Rokach, 2005). Tasks and methods are drawn from the areas of statistics, classification, clustering, visualization, and data mining (Rokach, 2005). Machine learning techniques (including deep learning) are also utilized.

Educational data mining techniques can be divided into two main categories: verification-oriented techniques, which rely on traditional statistical techniques such as hypothesis tests and analysis of variance, and discovery-oriented techniques that are used for prediction and categorization, such as classification, clustering, web mining, and others (Rokach, 2005). Those two categories may employ similar techniques, but for different purposes. For example, in the first case, a logistic regression model might be constructed with the aim of understanding its parameters (e.g., Kizilcec & Halawa, 2015), while in the second case the same modeling technique could be used for a purely predictive goal (e.g., Whitehill et al., 2015). Figure 3.1 below further breaks down the machine learning methods used for classification, which may also be used for other purposes.

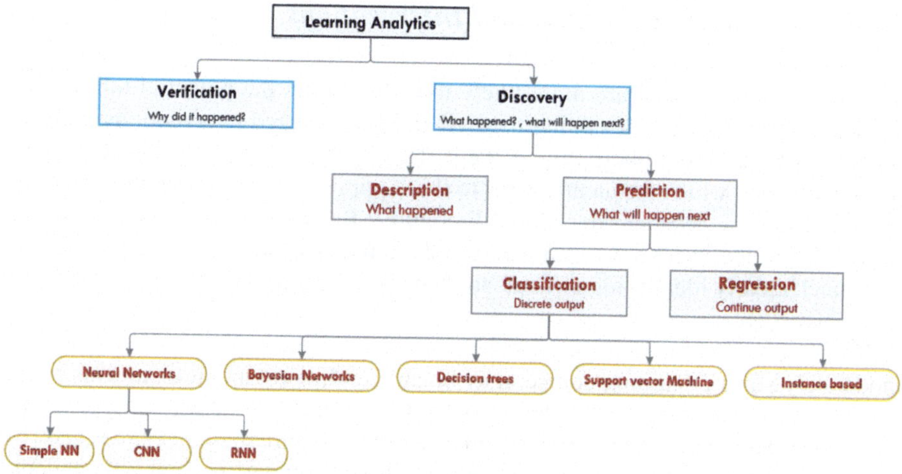

Fig. 3.1 Taxonomy of learning analytics for classification (adapted from Rokach, 2005)

Many research works have used features extracted from pure learning behaviour, as well as features related to social behaviour or interactions. Pure learning activities include visiting pages, watching videos, downloading files, and taking quizzes, while social interaction includes posting in forums, participating in discussions, sending private messages, participating in social networks, and so on. Data related to these activities may be obtained from sources like activity logs, data bases, and external sources to describe or predict outcomes like dropout, completion, success, and certification, using statistics, machine learning and deep learning algorithms. In the area of descriptive learning analytics, Cocea and Weibelzahl (2009) has proposed models to describe learner behaviour based on learning data. Many other projects have also led to the development of plug-ins used in Moodle for descriptive analysis (Mwalumbwe & Mtebe, 2017). Concerning predictive learning analytics, many authors have also produced valuable research works (Kotsiantis et al., 2013). There are also some plug-ins developed for predictive analysis (Mwalumbwe & Mtebe, 2017).

3.2.3 *Importance of Student Success Predictive Models in MOOCs*

In MOOCs, predicting student achievement is beneficial for a wide range of tasks. Many authors agree on three major reasons for constructing predictive models of student achievement, as described in the following subsections. Our model for the classification and prediction of learner behaviour in MOOCs addresses all three of these main purposes.

3.2.4 Personalized Support and Interventions

Identifying students who are more likely to or fail offers the potential to improve the student experience by enabling focused and personalized interventions to those students who are most likely to need help. This is the stated motivation for many previous works, which frequently refer to these pupils as "at-risk" students (a term adopted from the broader educational literature). Because of the large number of students enrolled in MOOCs compared to the amount of the instructional support personnel, clearly identifying difficult students is critical for delivering focused and timely assistance.

While a teacher in a regular in-person higher education course, or even a moderately sized e-learning course, may be able to directly monitor students, such observation is not possible to support MOOC instructors at scale, and predictive models can help with (a) identifying which students require these resources and (b) intervening by forecasting which resources will best support each at-risk student. Predictive models that can identify these individuals with high confidence and accuracy are necessary, especially when instructor time and resources are limited. Furthermore, many interventions would be superfluous or even harmful to the learning of students who are engaged or otherwise successful. A predictive model must generate accurate and actionable forecasts in order to provide individualized support and actions.

3.2.5 Adaptive Content and Learner Pathways

Predictive models in MOOCs have the potential to optimize the delivery of course content and experiences for projected student performance. In MOOCs, there has been very little research into adaptivity or true real-time intervention based on student success forecasts in any manner. For example, dropout prediction is used by Whitehill et al. to improve learner response to a post-course survey (this work optimizes for data collection, not learner performance) (Whitehill et al., 2015), and He et al. propose a hypothetical intervention based on projected dropout rates (but only implements the predictive model to support it, not the intervention itself) (He et al., 2015). Kotsiantis et al. offer a predictive model-based support tool for a 354-student distance learning degree programme, which is far smaller than most MOOCs. Pardos et al. implement a real-time adaptive content model in an edX MOOC. However, their approach is geared around increasing time spent on the page rather than improving student learning (Pardos et al., 2017).

To some extent, the paucity of research on adaptive content and learner pathways backed by accurate, actionable models at scale is owing to a lack of consensus on the most effective strategies for developing predictive models in MOOCs, which we address in the current study.

3.2.6 *Data Understanding*

Predictive models can be used as exploratory or explanatory tools, assisting in the understanding of the mechanisms underlying the desired outcome. Predictive models can also be used to detect learner behaviours, learner qualities, and course attributes related with MOOC performance, rather than just offering predictions to enable targeted interventions or adaptive material. These findings can help us enhance the content, pedagogy, and platform, as well as gain a better understanding of the underlying elements that affect student success in these settings. They also make a more direct contribution to theory by offering a better understanding of the complicated interactions between predictors and outcomes derived from the predictive modeling.

Certain types of models are more useful than others from this perspective. Models with simple interpretable parameters (like linear or generalized linear models, which provide interpretable coefficients and p-values, and decision trees, which generate human-readable decision rules) are far more useful for human understanding of the underlying relationship than models with many complex parameters (such as a multilayer neural network). Unfortunately, the latter are usually (but not always) more effective in making predictions in practice, so interpretability and predictive performance are frequently trade-offs. Some major developments in making increasingly sophisticated models interpretable suggest that this trade-off may be decreased in the future (e.g., Baehrens et al., 2009), but for predictive models in MOOCs, this "fidelity-interpretability trade-off" is still a major concern (Nagrecha et al., 2017).

3.2.7 *Common Metrics for Student Success in MOOCs*

Much research in the area of MOOCs has focused on analysing different learning outcomes including dropout, stopout, and certification, as well as measures such as final exam grade or final course grade. These outcomes are taken as measures of student success. Figure 3.1 shows the distribution of measures used in recent studies of success prediction (Gardner & Brooks, 2018).

The issue of evaluating student success in MOOCs is particularly problematic since indicators from traditional educational contexts—such as dropout, achievement, participation, and enrolment—can mean various things or appear illogical in the context of a MOOC. The authors frequently employ different definitions for these concepts, in the context of MOOCs. Furthermore, the abovementioned measures all focus on final outcomes and ignore the level of engagement of the student during the course. A balanced picture of student success should provide a broad collection of measures that assess course completion, engagement, and learning outcomes. Having various different measures to quantify MOOC effectiveness and outcomes allows us to test the robustness of models by potentially checking their

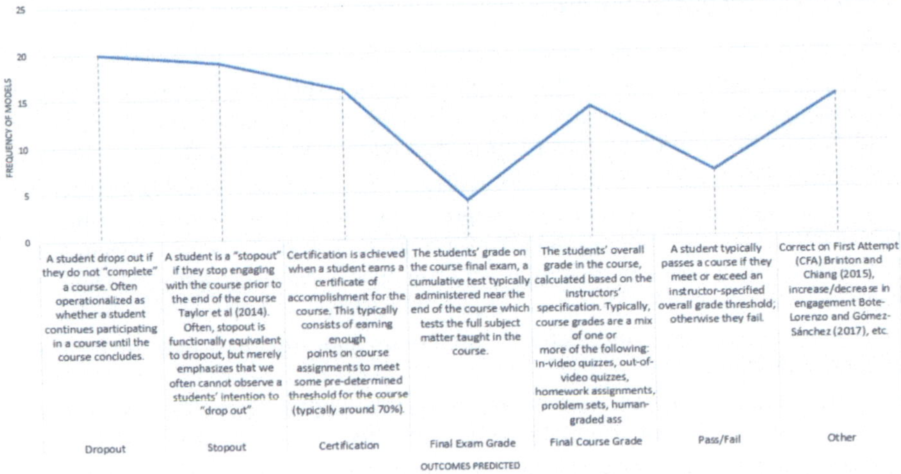

Fig. 3.2 Trend of predictive models according to outcomes predicted (adapted from Gardner & Brooks, 2018)

ability to predict multiple different outcomes. Furthermore, it allows us to capture metrics such as course completion, certification, and career advancement as shown in Fig. 3.2.

3.2.8 *Inputs Used by Student Success Predictive Models*

Besides the variety of outcomes predicted, there is also a variety of inputs used in MOOC predictive models. Those outcomes include logging status and frequency, attendance status, dropout status, completion, final grade, success, learning outcomes, learner behaviour, and so on. Table 3.1 below categorizes predictive models according to input types, with their associated data sources and usual outputs.

The figure below shows the tendency, in 2017, of predictive models in MOOC by types of inputs used.

3.2.9 *Activity-Based Models*

Activity-based models use inputs related to learner behaviour to evaluate learner behavioural outcomes such as dropout, failure, retention, success, and certification. From Fig. 3.3, it is evident that activity-based models are the most commonly found in the literature. This may be attributed to the fact that many outcomes predictable are more driven by activities. In addition, in MOOC platforms, activity

Table 3.1 Synthesis of predictive model categories from types of features perspectives

Predictive model categories	Type of input/features	Sources	Types of predictions/outcomes
Activity-based models	Use inputs data from learner behaviour to evaluate learner behavioural outcomes (dropout, failure, retention, success, certification)	Clickstream files data bases	Dropout, grade, certification
Demographics-based models	Use learner attributes which remain static (age, sex, academic level, town, marital status, country, town, . . .) over the interval of a course to predict student success	Data bases	Dropout, grade, certification
Learning-based models	Use observed student learning or performance on course assignments or theories of student learning as the basis for predictive modeling	Clickstream files data bases	Pass/fail, final grade, assignment, or exam prediction
Discussion forum and text-based models	Use natural language data generated by learners, as well as linguistic theory as the basis of student models	Clickstream-based activity features and natural language processing features	Completion
Cognitive models	Cognitive models incorporate observed or inferred cognitive states or rely on theories of cognition	Biometric tracking, contemporaneous questionnaires	Dropout, grade, certification
Social models	Social models use observed or inferred social relationships, or theories of social interaction, as the foundation for student models	For a data, external sources	Dropout, grade, certification

data are more abundant and more granular than any other data. Clickstream files, for example, provide detailed and granular interaction-level data on users' engagement or interaction with the platform.

In addition, Brinton and Chiang (2015) established that activity-based features not only predict activity-based outcomes but also appear to provide reasonable predictive performance even in non-activity-based prediction tasks, such as in grade prediction. Activity-based features may include simple counting-based features (like the number of posts in forum, the number of quizzes completed) (e.g., Xing et al., 2016), as well as more complex features such as temporal indicators of increase/decrease of course engagement (date of last connexion to the course

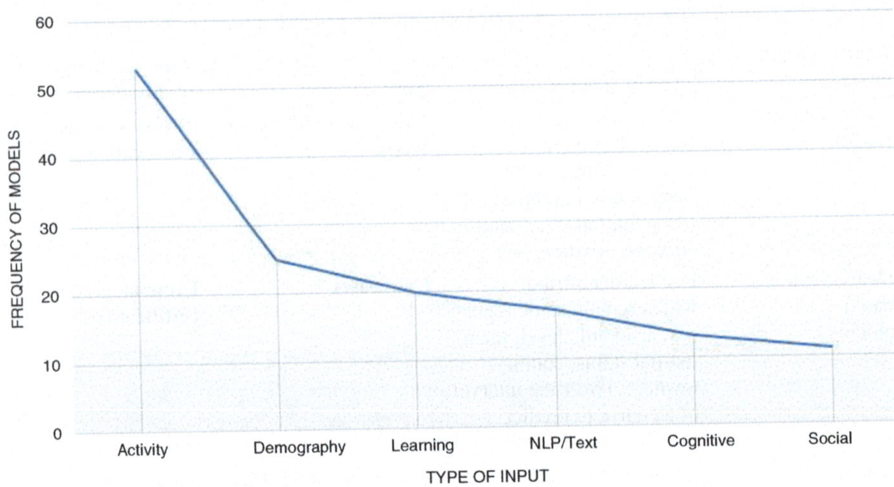

Fig. 3.3 Trend of predictive models in MOOC according to the types of inputs used (based on data from Gardner & Brooks, 2018)

platform) (Bote-Lorenzo & Gómez-Sánchez, 2017), sequences (Fei & Yeung, 2015), and latent variable models (Qiu et al., 2016). Despite the variety of types of activity-based inputs, all are obtained from the same underlying data source, namely the course clickstream log (which may be reformatted as a relational database consisting of extracted time-stamped clickstream events). Base features are drawn from a relatively small and consistent set of events, including page views, activity views, forum posts and views, and quiz completion: these features reflect the structure of courses available across the dominant MOOC platforms, such as edX and Coursera. Recently, however, due to the gamification of learning, different types of features have also emerged, as shown in Table 3.2 below.

3.2.10 Demographics-Based Models

Demographics-based models use learner attributes which remain static over the interval of a course to predict student success. Examples of demographic variables include age, sex, academic level, marital status, and so on. Several works have investigated the relationship between learner demographics and success in MOOCs. For instance, Qiu et al. (2016) examine the impact of both gender and level of education on forum posting, total active time, and certification rate for a sample of XuetangX MOOCs. They found that in non-science courses, females had higher rates of forum activities (posting and replying), more time spent on video and assignment activities and higher certification rates—but in science courses, the reverse was true.

Table 3.2 Synthesis of the common types of activity-based features

Types of activity-based features	Description	Example of work
Counting-based activity features	Predictive model, built on simple counting-based features extracted entirely from clickstream events	Predictive model, utilizing a support vector machine (Ramos & Yudko, 2008)
Early course activity features	Students' behaviour early in the course might be particularly predictive of their final performance in a MOOC. The fact that "early warning" systems anticipate based on limited amounts of data means that their projections can alter drastically throughout the early stages of a journey.	A simple logistic regression classifier based only on week 1 behaviour which effectively predicts certification in a MOOC offered to university students. Within the first two weeks of an e-learning history course, early access to course resources, such as a textbook and its integrated formative assessments, provides precise predictions of success or failure. Exercise in the first week of a nine-week course is a significant predictor of persistence in the course (Stein & Allione, 2014)
Temporal and sequential activity features	Rely on cumulative percentage of available lecture videos watched, the number of threads viewed on the forum, the number of posts made on the forum, the number of times the course progress page was checked), compiled over each week of the course. A particularly novel aspect of this work is the use of students' checking of their course progress page as an input feature.	Hidden Markov Model (HMM) to predict dropout. Students who never check their course progress have a dropout rate of 20–40% at each week of the course, while students who check their progress four or more times have a dropout rate of less than 5% each week. The use of an LSTM with sequential data is a promising technique, but it has to be replicated across a wider sample of courses (Fei & Yeung, 2015)
Latent activity features/variable	Because of its ability to infer complicated associations between variables in a data-driven manner, latent activity features have been widely used in predictive models of student achievement	To model student performance, probabilistic soft logic (PSL) is used to a set of activity- and natural language-based features. The latent variable assignments from this PSL method are used as predictors in a survival model (Kizilcec & Halawa, 2015)

(continued)

Table 3.2 (continued)

Types of activity-based features	Description	Example of work
Course metadata features	There are few researches into elements of courses (course metadata) that may be important to student activity inside the courses	Early engagement (registering earlier or completing a pre-course survey) is the strongest predictor of completion. Individual lecture titles are linked to different levels of engagement. Course subject interacts with learner demographics (i.e., gender) in predictive models (Qiu et al., 2016)
Higher order activity-based features	Other studies in MOOC research have starting to emerge, leveraging more complicated feature types. Explorations of higher order n-gram representations of learner activity data, which have shown promise predictive performance, are part of this	In activity-based n-gram models, counts of unique sequences of events or behaviours are used to create features, which are then used to build supervised learning models. Sequences of behaviour, irrespective of the time gaps between them, contain richer information than individual events or counts of these events without considering the context of other neighbouring events in time (Brooks et al., 2015)

Other works established links between demographic features like age, prior education, and prior experience with MOOCs and both dropout and achievement. Greene et al. (2015) identified those three features as significant predictors of both dropout and achievement. Khalil and Ebner (2014) identified several features that explained a large proportion of dropout in MOOC, including lack of time, lack of motivation, and "hidden costs" (textbooks needed for reference or paid certificates not clearly mentioned at the beginning).

On the other hand, Brooks et al. (2015) found that adding demographics provided only minimal improvement over the performance of activity-based predictive models for academic achievement of learners enrolled in MOOCs. Brooks et al. (2015) demonstrate that demographics-based models underperform activity-based models in MOOCs, even during early stages of the course when activity data is minimal. Additionally, demographic features provide no discernible improvement when added to activity-only models—on the contrary demographic features tend to degrade the performance of activity-only models in the second half of the course, as activity data accumulates (Brooks et al., 2015).

3.2.11 Learning-Based Models

Learning-based models use observed student performance on learning tasks (including course assignments) or theories of student learning as the basis for predictive modeling. Learning is obviously the basic objective of any MOOC—however, learning-based features are limited and are only used to predict a limited set of outcomes, such as pass/fail, final grade, assignment, or exam prediction. Bayesian Knowledge Tracing (BKT) (Mao, 2018; Pardos et al., 2013) has been widely used in intelligent tutoring systems to predict homework scores. However, Ren et al. (2016) found that "personalized linear regression" for predicting student quiz and homework grade outperformed an item-level variant of BKT (IDEM-KT) across two MOOCs.

Garman et al. (2010) applied pre-existing learning assessment to online courses by administering a commonly used reading comprehension test (the Cloze test) to students in an e-learning course. He found that reading comprehension is positively associated with exam performance and overall course grade but found no association between reading comprehension and open-book quizzes or projects. Kennedy et al. (2015) evaluated how prior knowledge and prior problem solving abilities predict student performance in a discrete optimization MOOC with relatively high prior knowledge requirements, drawing on robust learning theory results from in-person courses. The prior knowledge variables alone account for 83% of the variance in students' performance in this MOOC. The relationship between prior knowledge and student performance is well documented in traditional education research but is largely unexplored in MOOCs, despite the potential presence of many more students who lack prerequisite prior knowledge in MOOCs relative to traditional higher education courses.

Other works focusing on time-on-task and task engagement are also student performance concepts which have been applied extensively to educational contexts outside of MOOCs. Champaign et al. (2014) evaluate how learner time dedicated to various tasks within the MOOC platform (assignment problems, assessments, e-text, check point questions) correlates with their learning gain and skill improvement in two engineering MOOCs. They find negative correlations between time spent on a variety of instructional resources and both skill level and skill increase (i.e., improvement in students' individual rate of learning), using assessments calibrated according to Item Response Theory.

On the other hand, DeBoer and Breslow (2014) find that time spent on homework and labs in a Circuits and Electronics MOOC on edX predict higher achievement on assignments, while time spent on the discussion board or book is less predictive or not statistically significant. Additionally, time on the ungraded in-video quiz problems between lecture videos is found to be more predictive of achievement than time on lecture videos themselves.

Moreover, peer learning and peer assessment are also important theoretical concepts in education but have seen only limited applications in MOOCs to date. Ashenafi et al. (2016) examine models for student grade prediction which only use peer evaluation. They apply these models to traditional courses with web-based components but argue that their findings are also applicable to MOOC contexts. Peer assessment is used extensively in MOOCs, and its predictive capacity remains largely unexplored (Jordan, 2015).

3.2.12 Discussion Forum and Text-Based Models

Discussion forums are an embedded feature in every major MOOC platform and are widely used in most courses. A detailed analysis of the data from discussion forums provides the opportunity to study many dimensions of learner experience and engagement, which could not been identified elsewhere. This includes a rich set of linguistic features (derived from the analysis of the textual content of forum posts), social features (measured by the networks of posts and responses or actions such as likes and dislikes), and some behavioural features not available purely from the evaluation of clickstream data. Typically, discussion forum and text-based models use natural language processing (NLP) applied to data generated by learners, as well as linguistic theory as the basis of student models.

Crossley et al. (2016) compared the predictiveness of clickstream-based activity features and natural language processing features. They found that clickstream-based activity features are the strongest predictors of completion but discovered that NLP features were also predictive. Their work established that the addition of clickstream-based activity features improves the performance over a linguistic-only model by about 10%.

Other works (Tucker et al., 2014) found a moderate negative association between students' mood in posts regarding individual tasks and their performance on those assignments in an art MOOC. They also discover a little upward trend in forum post mood throughout the course of the study. Other predictive work in the area of sentiment analysis shows a link between sentiment and attrition that appears to be different depending on the course content (Wen et al., 2014).

3.2.13 Cognitive Models

As the foundation for student models, cognitive models incorporate observed or inferred cognitive states or rely on theories of cognition. Despite the fact that MOOCs are ultimately concerned with influencing learners' cognitive states (learning is a cognitive activity), there has been surprisingly little research on the use of cognitive data in MOOCs. This could be due to the particular difficulties of obtaining this data, especially when compared to other rich data sources (activity, forum postings, etc.). Novel data collection approaches, ranging from biometric tracking (e.g., Xiao et al., 2015) to contemporaneous questionnaires, are used in much of the research on cognitive states in MOOCs (Dillon et al., 2016).

Other authors like Wang et al. (2015) look at behaviours related to higher order thinking as displayed in student discourse and examine their relationship to learning using data from discussion forums. Several learning outcomes are evaluated using hand-coded data and a learning activity classification framework based on cognitive science research. The authors discovered that students who used "active" and "constructive" behaviours in the discussion forum, behaviours that demonstrate higher level cognitive tasks like synthesis rather than simply paraphrasing or defining, produced significantly more learning gains than students who did not use these behaviours. Furthermore, they established that useful cognitive data relevant to student performance may be retrieved from discussion forum posts and applied to simple models using techniques such as bag of words and linear regression. Furthermore, cognitive strategies if they can be discovered properly appear to be associated with student performance in MOOCs and that cognitive theory can be used to inform MOOC prediction models.

Novel data collection approaches are used in much of the work in this subject. Future research should move increasingly beyond questionnaires and self-reports as the sole source of cognitive data from learners. The type of data required for this type of research should become more available for researchers as sensing technology gets more affordable and consumers' devices (such as smartphones and tablets) become increasingly outfitted with sensors. Many canonical cognitive findings in educational research have yet to be investigated or replicated in a MOOC environment, and further study is needed to establish the limitations of these findings when applied to MOOCs.

3.2.14 Social Models

Social models of learning are built on the foundation of observed or assumed social relationships or ideas of social interaction.

Many studies employ discussion forums to build social networks in which students serve as nodes and varied response relationships serve as edges. Joksimović et al. (2016), for example, use two sessions of a programming MOOC, one in English and one in Spanish, to assess the association between social network ties and performance (specifically, non-completion vs. completion vs. completion with distinction). Students who received a certificate or distinction were more likely to interact with one another than non-completers. Jiang et al. (2014) discovered that learners in different performance groups tend to communicate with one another in different types of MOOCs. Joksimović et al. (2016) found that weighted degree centrality was a statistically significant predictor of completion with distinction in the both the English and Spanish courses mentioned above, as well as a significant predictor of basic completion in the Spanish language course. On the other hand, closeness and betweenness centrality had more variable and inconsistent effects across courses. They get to the conclusion that structural centrality in the network is related to course completion (Joksimović et al., 2016). The findings are similar to those of Russo and Koesten (2005), who found centrality to be a statistically significant predictor of student achievement in a short online course. The findings are similar to those of Russo and Koesten (2005), who found centrality to be a statistically significant predictor of student achievement in a short online course.

In a related study, Dowell et al. (2015) examine how text discourse features can predict social centrality and that discourse features explain about 10% of the variance in performance (compared to 92% with a model using discourse + participant features). The explained variance increased to 23% for the most active participants in the forums.

There is a need for more research into the impact of social networks in MOOCs, as well as more exploration of external social network data. Although social networks appear to play a major role in students' learning, they are difficult to quantify using existing MOOC data, especially with small single-course samples. Despite the richness of these data sources, MOOCs rarely incorporate external digital social networks (such as data from Facebook or LinkedIn). Existing research, on the other hand, appears to be unduly reliant on discussion forums as sources of social network data. The evaluation of new data sources on social issues has the potential to have a significant impact on the scholarly consensus in this field.

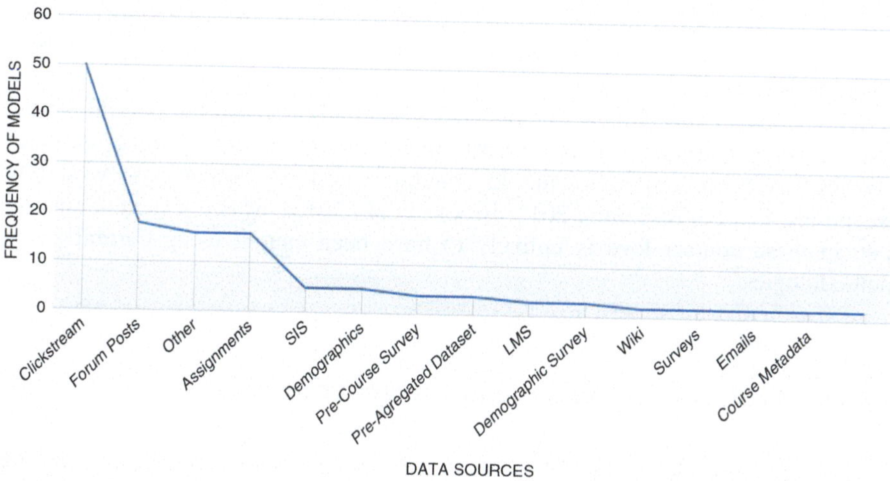

Fig. 3.4 Trend of predictive models in MOOC according to the data sources providing inputs used (based on data from Gardner & Brooks, 2018)

3.2.15 Data Sources Providing Inputs/Features for Predictive Models

The data sources employed in MOOC predictive modeling research have received little attention. Knowing which data sources are valuable for prediction and which are unexplored is a good starting point for future research. Recognizing which data sources are most valuable can also increase the efficiency of predictive modeling work in practice because feature extraction is costly in terms of both development and computing time.

The figure below displays the common data sources used in predictive models designing in MOOCs. Moreover, this figure confirms that clickstreams are the most common raw data source for predictive modeling research in MOOCs, out of the raw data sources outlined in Fig. 3.4 above.

In some ways, the much used of clickstream data is unsurprising: clickstreams give rich granular data that the field is only just beginning to understand how to capture in its entirety. Clickstreams, on the other hand, are unstructured text files that need a lot of human and computational work to parse. Due to faults in platform server logging, their forms are complex and sometimes inconsistent, and a single item can have multiple levels of aggregate applied to it. The various data types in the figure above are usually given as organized relational databases that may be accessed using simple SQL commands. The fact that clickstreams are so frequently used, despite the difficulties in acquiring and using this data, demonstrates their utility in predictive modeling. The authors (Gardner & Brooks, 2018) compare the predictiveness of clickstream features vs. forum- and assignment-based features when predicting dropout across the entire population of learners in a large state university in the USA. This work verifies that clickstream features are more effective predictors than forum- or assignment-based features when predicting dropout across the entire population of learners.

While clickstreams contain complicated, potentially relevant temporal information regarding learner behaviour across time, most modeling of these temporal patterns has been limited to simple counting-based representations (with few exceptions; i.e., Fei & Yeung, 2015; Brooks et al., 2015). Much of the complexity seen in these contact logs is unlikely to have been caught using current study methodologies.

3.2.16 Features Engineering in Predictive Models

Boyer and Veeramachaneni (2015) show how good feature engineering may be paired with effective statistical models to produce performant student success predictors in a series of papers. Several unique approaches to developing activity-based models of student success in MOOCs are demonstrated in these studies, which combine crowd-sourced feature extraction, automatic model tweaking, and transfer learning.

Boyer and Veeramachaneni (2015) employ crowd-sourced feature extraction to create behavioural features for stopout prediction, using members of a MOOC to apply their human skills and domain knowledge. For all four cohorts studied, the authors find that these crowd-proposed characteristics are more complex and have greater predictive performance than simpler author-proposed features (passive collaborator, wiki contributor, forum contributor, and fully collaborative). The predictive model in this study is based on a basic regularized logistic regression, revealing that many good predictive models of student achievement in MOOCs have depended on creative feature engineering rather than complicated algorithms.

Boyer and Veeramachaneni conclude that a posteriori models, which are built retrospectively using labeled data from the target course, provide an "optimistic estimate" and "struggle to achieve the same performance when transferred." The same researchers discovered that an in situ prediction architecture transfers well, with performance comparable to a model that takes into account a user's whole history (which is not actually possible to obtain during an in-progress course).

The two sessions above helped to explore the data sources commonly used and to address the issue of extracting features from those sources. It appears that clickstreams are the most common raw data source used for predictive modeling research in MOOCs. In the session above, we saw that many types of activity-based features are embedded in clickstreams. Activity-based features are the basics of activity-based models that are the most commonly used predictive models. From the explanations provided above, it appears that the model we are developing for the classification and prediction of learner behaviours is an activity-based model.

3.2.17 Relation Between Types of Model and the Outcome Predicted

Above we presented various types of models, ranging from activity-based to social model. We also presented the most common types of outcomes that have been predicted in MOOCs, and they include dropout, completion, and others.

The figure below is adapted (Gardner & Brooks, 2018), and the results presented were established by experiments that included a predictive model that could be classified as many categories or predicted numerous outcomes were included in each category in this table, resulting in cell totals that surpass the total number of works assessed. Pass/fail, final grade, assignment grade, and exam grade are examples of "academic" outcomes.

All measures of course completion, such as certification and participation in the final course module, are included in the term "complete." They established that academic outcomes are the most predicted, while completion and other types of outcomes are the least predicted. On the other hand, while activity-based model is the most common, cognitive-based models have not been well explored.

The figure below groups several outcomes into a single "academic" outcome category. "Pass/Fail" indicates whether a learner met a predetermined final grade threshold to pass the course, and "Certification/Completion" indicates whether a student successfully completed all course requirements and received an official certificate of completion. In addition, those official certificate sometimes requires payment and identity verification; Fig. 3.5 shows the various types of models according to the outcomes predicted.

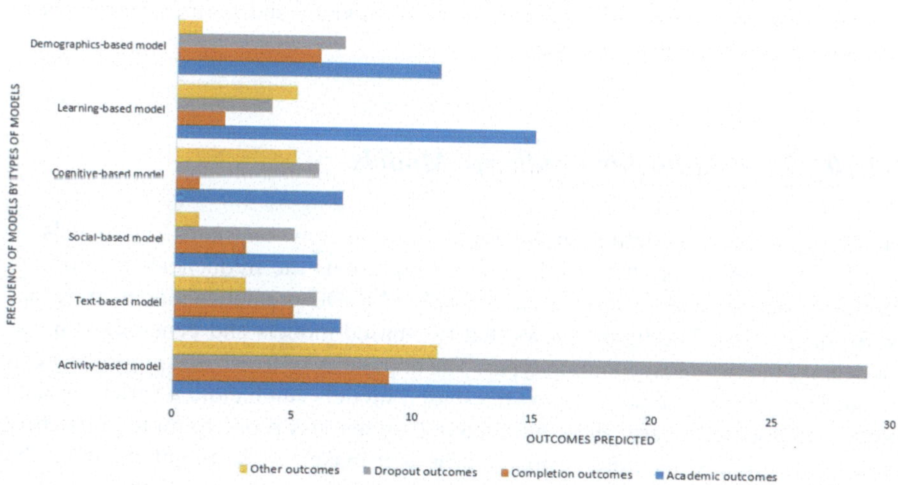

Fig. 3.5 Trend of types of predictive models (inputs used) in MOOC according to the outcomes predicted (adapted from Gardner & Brooks, 2018)

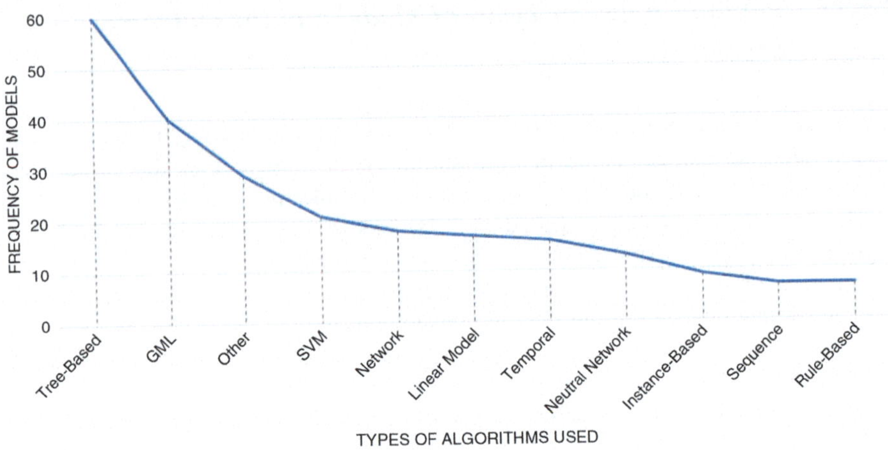

Fig. 3.6 Trend of predictive models in MOOC according to the types of algorithms used (adapted from Gardner & Brooks, 2018)

3.2.18 Algorithms for Predictive Models and Metrics for Their Evaluation

Once the outcomes to be predicted are identify, as well as the type of model, the features and the sources to extract the features from, the remaining tasks include selecting relevant algorithms to build/train the model and the suitable metrics for its evaluation. According to outcomes predicted and the model types, a number of algorithms have been commonly used to build models, and various metrics have been used to evaluate those models.

3.2.19 Algorithms for Predictive Models

Predictive student modeling in MOOCs relies heavily on statistical models to translate features to predictions. Figure 3.6 represents the frequencies of different classes of statistical algorithms used to develop MOOC predictive models (Gardner & Brooks, 2018). The figure shows that tree-based models and generalized linear models are the two most common types. The popularity of tree-based algorithms can be attributed to several advantages: tree-based models can handle a variety of data types (categorical, binary, and continuous), they are less prone to multicollinearity than linear models, they are nonparametric and make few assumptions about the underlying data, and their outputs may be interpreted through visualization, inspection of decision rules, variable importance metrics, and other methods. On the other hand, GLMs are quick and easy to fit to data, requiring little or no hyperparameter tuning. Unlike tree-based models, they produce regression coefficients that can be

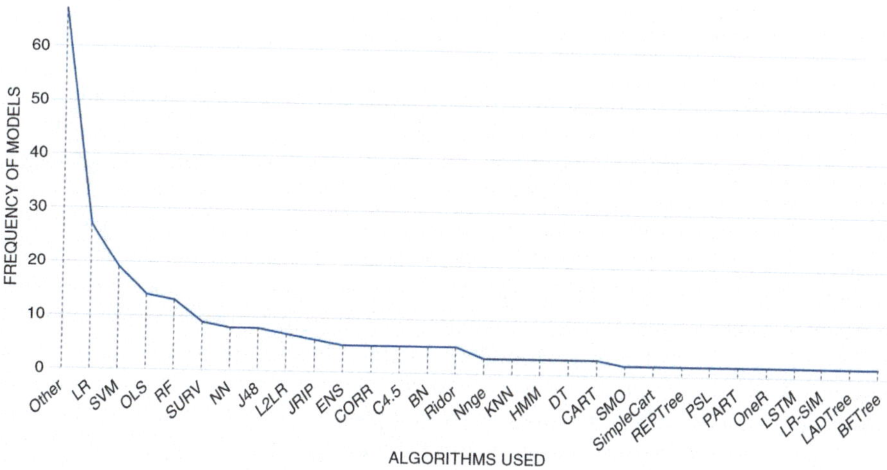

Fig. 3.7 Trend of predictive models in MOOCs according to the specific algorithms used (adapted from Gardner & Brooks, 2018)

directly interpreted as expressing the relative contribution to overall accuracy of the different predictors.

Figure 3.7 further disaggregates Fig. 3.6 by showing the specific algorithms utilized. The figure shows that how the prevalence of tree-based algorithms obscures the lack of uniformity in the algorithms utilized. It appears that, of all the tree-based algorithms studied, only random forests were used in more than ten of the works. As a result, evaluating the effectiveness of any particular tree method across their survey is challenging. In contrast, there are few GLM algorithms used in the literature; practically, all GLM algorithms are logistic regression (LR) and L2-penalized logistic regression ("ridge" regression, L2LR). Despite their high parametric assumptions about the underlying data, GLMs, and L2LR in particular, often achieve outstanding performance when applied with large and robust feature sets.

Finally, Fig. 3.7 shows a "long tail" of modeling methodologies, with about half of the work using customized, individualized algorithms, indicated by "Other" in Figs. 3.6 and 3.7. This reflects a focus on innovation in academic research, as well as a new area with limited consensus on the optimal strategy to solve prediction problems. We notice that none of the methods in the assessed work consistently outperforms all other algorithms, implying that there is no one "best" algorithm for a given job or dataset (Wolpert & Macready, 1997). At this juncture, future work comparing and evaluating the fitness of various predictive modeling algorithms for various objectives in MOOC research would be suitable. It also appears that supervised learning approaches dominate the literature, with few examples of unsupervised approaches; this is likely due to the fact that many of the outcomes (i.e., dropout, certification, pass/fail, grades) are observable for all learners, making unsupervised techniques unnecessary for many of the prediction tasks addressed by research to date.

3.2.20 Metrics for Model Evaluation

The figure below shows the distribution of evaluation measures commonly used as highlighted in Gardner and Brooks (2018). There is a general consensus on a set of evaluation metrics, including accuracy (ACC), area under the Receiver Operating Characteristic curve (AUC), precision (also known as positive predictive value) (PREC), recall (REC) (also called true positive rate, sensitivity, or probability of detection), F1, and kappa. Different measures evaluate different aspects of predictive quality, which change based on the task and study goals. However, sometimes readers are unable to compare performance across otherwise-similar studies that report different performance metrics due to the lack of a common baseline. Reporting multiple measures would frequently provide a fuller view of model performance and make cross-study comparisons easier, while still allowing researchers to look at performance using their preferred metric(s). Open data or open replication frameworks would enable more detailed comparisons and shift the burden of proof from the researcher to reviewers and critical readers, who would be able to evaluate results using any performance indicator of interest.

Classification accuracy is reported as the only model performance metric in ten of the studies surveyed (more than 10%). But although classification accuracy is easily interpretable, but it can be a misleading measure of prediction quality when outcome classes are of unequal size, as is the usual case in MOOCs (i.e., most students dropout, do not certify, etc.). The same data that is used to compute accuracy can also be used to obtain more informative performance metrics, such as sensitivity, specificity, F1, Fleiss' kappa, and so on. Other measures, such as the AUC, assess performance for all potential thresholds, thus taking into account the fact that performance is threshold dependent, as shown in Fig. 3.8 below.

The best model evaluation metric is typically determined by the outcome being measured as well as the specific objectives of a predictive modeling project. For example, recall may be an appropriate model evaluation metric in a dropout modeling experiment where the goal is to provide an inexpensive and simple intervention to learners (such as a reminder or encouragement); however, precision may be a better choice when the goal is to provide an expensive or resource-intensive intervention to predicted dropouts.

3.2.21 Lessons Learned from Related Work

The accuracy dimension of predictive student models described in the figure above is reflected by the data source, feature extraction method, statistical modeling algorithm, and assessment metric taken together. Research into and methodological development in each of these domains (feature extraction, modeling, and evaluation) stands to significantly increase the accuracy of future predictive MOOC models.

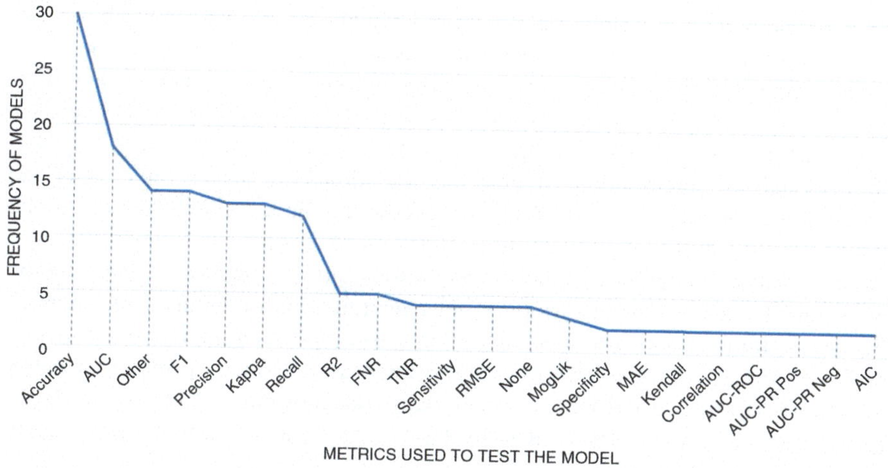

Fig. 3.8 Trend of predictive models in MOOC according to the evaluation metrics used (adapted from Gardner & Brooks, 2018)

We have observed the various aspects of the related works and the key observations made are the following: (1) input/features/predictors have been focusing most on exploring activity-based data through click stream logs. (2) Outcomes predicted or the prediction have been focusing on (a) dropout, (b) completion, (c) success, and (d) certification, not providing the actor with a good understanding of what is going on in the MOOC platform which remain a "Black Box," Course instructor is not provider with a tool clear understanding of the learner behaviour in order to support him in the learning process. (3) Algorithm models have been focusing most on the descriptive models; in addition, predictive models used so far have not explore enough the power of deep learning for better prediction.

Given the previous observations, we conclude that the following actions would provide actors with tools for better management of the learning process in MOOCs: (1) exploring more features, (2) exploiting the power of deep learning, and (3) classifying and predicting learner behaviour.

3.2.22 Approach

Traditional educational researchers and practitioners used methods such as (1) surveys, (2) interviews, and (3) observations, those methods are (1) time consuming, (2) costly, and (3) do not provide the course instructor with timely and useful information to understand and manage the teaching and learning process, and in addition, a delay always occurs between data collection and prediction. So educational researchers progressively switched to learning analytics (LA) for real-time analysis of data generated by the learning process.

LA includes both descriptive and predictive analyses. As described above, our goal is to construct a predictive model that takes into account a variety of features of a learner life cycle (activity-based features, demographics-based features, learning-based features, Discussion Forum and Text-Based Features, Cognitive Features, Social Features) in a MOOC platform, for the purpose of enabling timely, targeted, and personalized intervention to promote retention and successful course and programme completion, by classifying and predicting the learning behaviour in the MOOC. For this purpose, it is necessary to take into account the characteristics of data to be analysed such as volume, variability, velocity, veracity, as well as the highly imbalanced nature of dropping out over retention or failure over success. The large volume of data available, as well as the complicated interaction of factors involved, indicates that deep learning algorithms can potentially provide far better prediction performance and accuracy than traditional algorithms (Xing & Du, 2019). The experimentation will start by activity-based features and then will progressively consider other types of features.

Our overall methodology consists of (1) exploring the potential of deep learning techniques to provide learner behaviour predictive model which can potentially outperform the traditional-used Machine Learning approaches, (2) analysing personalized interventions using individual's learner behaviour prediction, and (3) checking whether deep learning models can better personalized and prioritised interventions to support learner in the learning process than other algorithms. In the current chapter, we deal exclusively with point (1). In order to achieve predictions that are actionable in a real-world context, and given time series types of data generated by learner behaviour, the training technique is L2 regularization on models using RNN architectures (Che et al., 2018).

While training our model, we initially play on (1) the number of epochs, (2) the type of RNN, (3) the number of hidden layers, (4) the learning rate, and (5) the regularization parameter. Accuracy is the most suitable model evaluation metric to work, in view of the binary nature of our outputs and the dataset. So, we choose accuracy as the metric to evaluate the model. During the training, we keep constant the parameters related to the Adams optimizer, as shown in Fig. 3.9 below.

3.2.23 Context/Dataset

The context for this work is a gender-sensitive STEM Education course deployed on Moodle. The course took place in September 2018 and lasted for 6 weeks. The course consisted of 6 modules and 3617 students registered. The course consisted of involved many activities and resources, including forums, quizzes, assignments, videos, audios, wiki, downloadable files, lecture content pages, announcement, calendar, and gradebook, whereas Moodle has many data sources, for this work our dataset will be extracted from clickstream logs, which is a file provided directly by Moodle containing historical information about the learners' and instructors' interactions in the platform, such as pages visited including when and how many

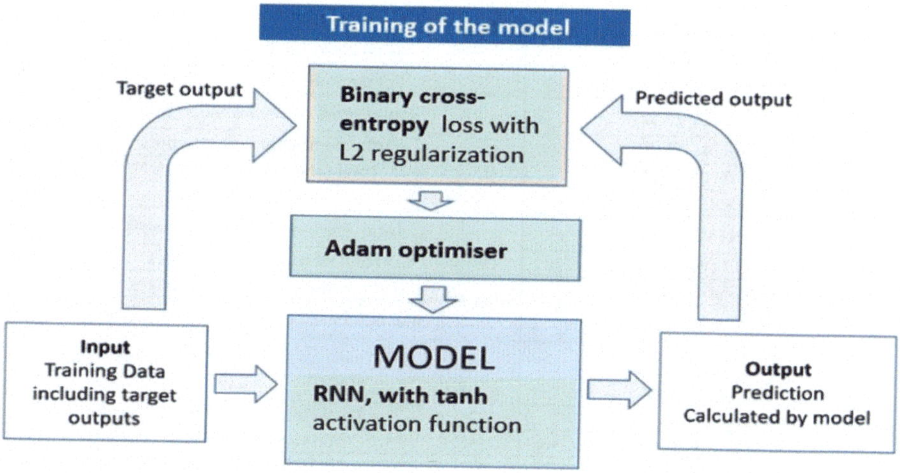

Fig. 3.9 Architecture of the training of the model

students and teachers interacted with the course content. The dataset also includes demographic data. Those sources are enough for the current preliminary study, because we do not intend at this level to go deep in the contents of learners' interactions.

3.2.24 Method

3.2.24.1 Features Generation

In order to build our predictive model, several activity-based features related to learners' interaction with the learning process or course content (activities and resources) were generated. The features and their descriptions are given in Table 3.1. These are all activity-based features and were chosen according to the MOOC prediction literature (Rosé et al., 2014) and our previous work (Sinha et al., 2014). Since feature engineering is not a principal objective of this work, we used a flat feature structure containing clickstream, directly provided by the log file. Our work adopts a classical 80/20 train/test split because we did not have a very large amount of data, as shown in Figs. 3.10 and 3.11 below.

3.2.25 Building the Model

Our model works on time series data (learner behaviour in the learning process), so RNN (Recurrent Neural Network) is a suitable deep learning algorithm to use

Feature	Description
TTL_DISC_VIEW	Total views on discussion forums
TTL_DISCPOSTCOUNT	Total number of discussion posts
TTL_VIDEO_VIEW	Total views on review-video pages
TTL_QUIZATTEMPT	Total number of quiz attempts
TTL_QUIZTIMESPENT	Total time spent on quizzes
TTL_QUIZSCORE	Total quiz score
TTL_INTRO_VIEW	Total views on introduction pages
TTL_LECCONT_VIEW	Total views on lecture pages
TTL_ASSIGN_VIEW	Total views on assignment pages
TTL_ASSIGN_SBM	Total number of assignment submissions
TTL_ANNOUN_VIEW	Total view on announcement pages
TTL_WIKI_VIEW	Total view on wiki pages
TTL_GRADEBOOK_VIEW	Total view on gradebook
TTL_CALENDAR_VIEW	Total view on calendar
TTL_AUDIO_VIEW	Total view on audios
TTL_FILE_VIEW	Total view on files
TTL_ACTIVITY	Total number of activity taken by the learner
LAST_ACTIVITY	Last activity taken by the learner
TTL_DAY_LAST_ACTIVITY	Total number of days since the learner took the last activity
....	....

Fig. 3.10 Overview of features generated to serve as inputs for the predictive mode

Page: 1 2 3 4 5 6 (Next)

Time	IP address	User full name	Action	Information
Wed 8 January 2020, 6:50 PM	196.207.229.154	José Meku	forum view forum	Forum introductif / présentation
Wed 8 January 2020, 6:50 PM	196.207.229.154	josé Meku	forum view forum	Forum introductif / présentation
Wed 8 January 2020, 6:50 PM	196.207.229.154	José Meku	forum add post	Re: Forum introductif / présentation
Wed 8 January 2020, 6:49 PM	196.207.229.154	josé Meku	forum view forum	Forum introductif / présentation
Wed 3 July 2019, 2:45 PM	196.183.38.195	Célestine BACON	forum view forum	Forum introductif / présentation
Mon 17 December 2018, 3:51 PM	41.203.147.4	Idrissa MOUSSA	forum view forum	Forum introductif / présentation
Sun 9 December 2018, 9:48 PM	154.68.5.144	Akoua Desire KOUAME	forum view forum	Forum introductif / présentation
Sun 2 December 2018, 10:47 PM	41.219.50.234	Alimatou Absa DIARRA	forum view forum	Forum introductif / présentation

Fig. 3.11 Example of clickstream data extracted from the Moodle platform/clickstream data having 8844 records

(Che et al., 2018). The most common RNN architectures used in predictive models are simple RNNs, GRUs, and LSTMs. All the RNN models work with sequenced data and feed information about previous states or time steps into each next state. Practically, LSTMs require the most memory, followed by GRUs. Simple RNNs have the smallest memory capacity (Anani & Samarabandu, 2018).

This chapter implements all the three aforementioned architectures. The model consists of a recurrent layer with 200 hidden layers coupled with a tanh activation function, and then the output is given to a layer, having a sigmoid activation function to produce the probability that the learner will take the next activity in the learning process. This probability will be used to classify and predict their behaviour. The model is built in such a flexible way to allow easy switching from one RNN architecture to another. This enables us to compare the models built using the three types of RNNs. The model also makes it possible to adapt the dimension of the RNNs to suit the input dataset.

In our experimentation, some learner's activities or interactions with the learning environment have been identified, and separate feature values are defined for each of those activities.

Learner's activities are related to the various tasks of a learner in an online learning environment. Those tasks include and are not limited to: (1) Log into the platform, (2) Using forums, (3) Using quiz. If we consider learners activities associated with the task (1) Log into the platform, then the features we can use, concerning learners, are "never log in", "log in on time". So for this activity, the model classifies and predicts learners who could log into the platform or not, as well as those who could log into the platform on time or not. If we consider learners activities associated with the task (2) Using forums, then the features we can use, concerning learners, are "view forum messages", "post messages in forum". So for this activity, the model classifies and predicts learners who could view forum messages or not, as well as those who could post messages in forum or not. If we consider learners activities associated with the task (3) Using quiz, then the features we can use, concerning learners, are "viewing quizzes", "completing quizzes", "validating quizzes". So for this activity, the model classifies and predicts learners who could view quizzes or not, those who could complete quizzes or not, as well as those who could validate quizzes or not...

- Training algorithm: ADAM optimizer
- ADAM uses the running averages of the previous gradients to adjust parameter updates during the training phase
- For the cost function J, represented below, we use binary cross-entropy loss with L2 regularization (helps model to fit the training data well and generalize better)

This cost function is constructed based on binary cross-entropy loss coupled to L2 regularization. Our model is performing a binary classification (learner takes the next activity or not) with estimated probabilities, compared to currently used training techniques like cross-validation, in situ or gradient clipping, so L2 regularization is appropriated because it helps smoothing oscillations in the training loss. The Adam optimization algorithm was used to train the model; Adam is an optimizing algorithm that uses the running averages of the previous gradients to adjust the parameters of the model during the training phase, and we play on (1) the number of epochs, (2) the type of RNNs, (3) the number of hidden layers, (4) the learning rate, and (5) the L2 regularization parameter. During the training, we keep constant the parameters related to the Adams optimizer. In addition, the learning

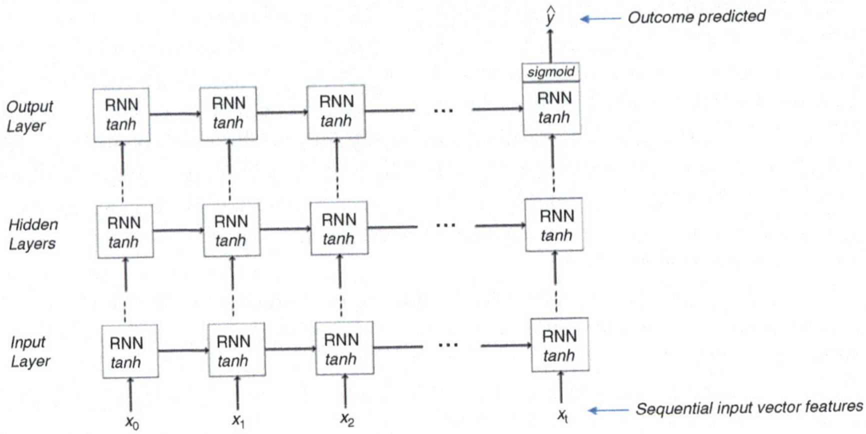

Fig. 3.12 Structure of the model with hidden layers

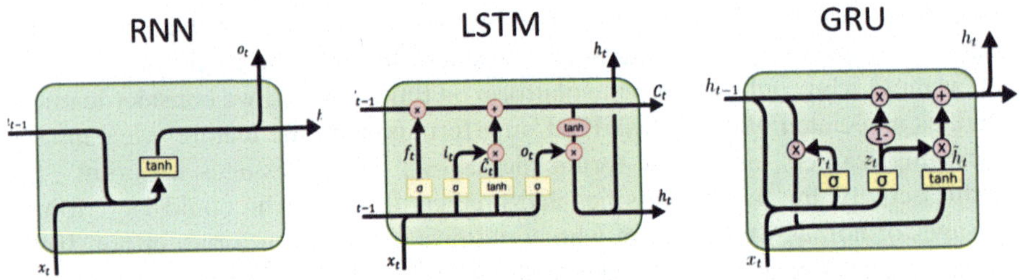

Fig. 3.13 Structure of any single type of RNN tested (simple RNN, LSTM, and GRUs)

$$J = -\frac{1}{m}\sum_{i=1}^{m}(ylog(\hat{y}) + (1-y)log(1-\hat{y})) + \frac{\lambda}{2m}\sum_{l=1}^{L}||w^{[l]}||_F^2,$$

- ✓ λ = L2 regularization parameter
- ✓ m= number of inputs
- ✓ W[l]= the weight matrix for the lth layer
- ✓ L= number of layers

Fig. 3.14 Cost function: binary cross-entropy loss coupled to L2 regularization

rate used by the model is automatically reduced during the training, if the validation loss did not decrease after a few numbers of epochs, as shown in Figs. 3.12, 3.13 and 3.14 above.

Table 3.3 Comparison of model accuracy for different types of the RNNs used

Inputs type	Simple RNN	GRU	LSTM
All features	0.8920	0.8801	0.8870

3.3 Experimental Results and Discussion

We experimented the model with many sets of hyper parameters, by try and error, with the three RNN architectures (simple RNNs, GRUs, and LSTMs) before concluding that simple RNNs perform best, as shown in Table 3.3, using a regularization parameter of $\lambda = 0.01$, which produced the best accuracy of 89.2% for simple RNNs.

With any of the three architectures, 200 hidden layers appear to be offering the best balance in terms of speed and accuracy compared to models with 64 or 256 hidden layers. In our work, we found that simple RNNs produced the best accuracy for the model on the dataset used. This was not expected as experience and previous works tend to predict that LSTMs should perform better than other types of RNNs. One potential explanation of this situation is that having long memory seems not to be so important in this dataset as expected. In fact, some previous works suggest that there might be a transition point in the learning process where learner behaviour far likely changes. Furthermore, past studies on learner behaviour suggest that learner's activities are high at the beginning of the learning process, then decrease during the process, and then slightly increase at the end of the process. The previous explanations imply that there is not a real need of advanced memory capacities for the required model, and then LSTMs may not necessarily be the best option.

3.4 Conclusion and Future Work

Our main objective in this chapter was to study relevant algorithms for the development of deep leaning model to classify and predict learner's behaviour in MOOC. Given the time series type of dataset, we tested three architectures of RNNs to find out that simple RNN with input features offers the best precision (performance and accuracy) in classifying and predicting learning behaviour in the learning process. One of the key benefits of this model is the fact that, by giving a good understand of learner's behaviour, the model might guide teachers to provide personalized support and interventions to learners in the learning process. This would give a tool to the course instructor/teacher who is the main tutor of any learning process, who mastered the course content, and who can better assist learners so that they actively benefit from the entire course. This analysis also reveals that learner's behaviours concerning video viewing, posting in discussion forums, viewing discussion forums and quizzes could help predict learner behaviour about other types of content. This model could also be used to support adaptive content and learner pathways, by suggesting the revision or restructuring of the content and/or training path, closure of a course, or the launch of a new course. This model could also support data understanding, by providing insight information, exploratory or explanatory tools, assisting in the understanding of the mechanisms underlying the desired outcome, and then helping to understand the mechanisms behind the outcomes.

The main challenge of this study was the imbalanced nature of our dataset in the single MOOC used for this first experimentation. In addition, the learner behaviour or interaction with a given content is not easily and deeply measurable. We used common behaviours like clicking, viewing, downloading, uploading, attempting, and posting which are not always suitable and easy to measure.

Future research would address the following main issues: (a) develop a method to measure the quality of learner behaviours, (b) test the model in other MOOCs and explore methods to further improve the deep learning behaviour prediction model performance in MOOCs by increasing the hidden layers in the network, (c) the current study only shows statistical validity of the model, (d) further researchers could examine its validity by implementing the model in ongoing MOOC courses to assess it in real-time prediction, and (e) subsequent research should also include designing personalized interventions based on the model predictions.

References

AFRICAINE, U. (2016). Stratégie continentale de l'education pour l'afrique.

Anani, W., & Samarabandu, J. (2018). Comparison of recurrent neural network algorithms for intrusion detection based on predicting packet sequences. In *2018 IEEE Canadian Conference on Electrical & Computer Engineering (CCECE)* (pp. 1–4). IEEE.

Andres, J. M. L., Baker, R. S., Gašević, D., Siemens, G., Crossley, S. A., & Joksimović, S. (2018). Studying MOOC completion at scale using the MOOC replication framework. In *Proceedings of the 8th International Conference on Learning Analytics and Knowledge* (pp. 71–78).

Ashenafi, M. M., Ronchetti, M., & Riccardi, G. (2016). Predicting student progress from peer-assessment data. In *International Educational Data Mining Society*.

Baehrens, D., Schroeter, T., Harmeling, S., Kawanabe, M., Hansen, K., & Müller, K.-R. (2009). How to explain individual classification decisions. Preprint, arXiv:0912.1128.

Bote-Lorenzo, M. L., & Gómez-Sánchez, E. (2017). Predicting the decrease of engagement indicators in a MOOC. In *Proceedings of the Seventh international Learning Analytics & Knowledge Conference* (pp. 143–147).

Boyer, S., & Veeramachaneni, K. (2015). Transfer learning for predictive models in massive open online courses. In *International Conference on Artificial Intelligence in Education* (pp. 54–63). Springer.

Brinton, C. G., & Chiang, M. (2015). MOOC performance prediction via clickstream data and social learning networks. In *2015 IEEE Conference on Computer Communications (INFOCOM)* (pp. 2299-2307). IEEE.

Brooks, C., Thompson, C., & Teasley, S. (2015). A time series interaction analysis method for building predictive models of learners using log data. In *Proceedings of the Fifth International Conference on Learning Analytics and Knowledge* (pp. 126–135).

Brooks, C., Thompson, C., & Teasley, S. (2015). Who you are or what you do: Comparing the predictive power of demographics vs. activity patterns in massive open online courses

(MOOCs). In *Proceedings of the Second (2015) ACM Conference on Learning@ Scale* (pp. 245–248).

Brown, R., Lynch, C. F., Wang, Y., Eagle, M., Albert, J., Barnes, T., Baker, R. S., Bergner, Y., & McNamara, D. S. (2015). Communities of performance & communities of preference. In *EDM (Workshops)*. Citeseer.

Champaign, J., Colvin, K. F., Liu, A., Fredericks, C., Seaton, D., & Pritchard, D. E. (2014). Correlating skill and improvement in 2 MOOCs with a student's time on tasks. In *Proceedings of the First ACM Conference on Learning@ Scale Conference* (pp. 11–20).

Che, Z., Purushotham, S., Cho, K., Sontag, D., & Liu, Y. (2018). Recurrent neural networks for multivariate time series with missing values. *Scientific Reports, 8*(1), 1–12.

Cocea, M., & Weibelzahl, S. (2009). Log file analysis for disengagement detection in e-learning environments. *User Modeling and User-Adapted Interaction, 19*(4), 341–385.

Crossley, S., Paquette, L., Dascalu, M., McNamara, D. S., & Baker, R. S. (2016). Combining click-stream data with NLP tools to better understand MOOC completion. In *Proceedings of the Sixth International Conference on Learning Analytics & Knowledge* (pp. 6–14).

DeBoer, J., & Breslow, L. (2014). Tracking progress: Predictors of students' weekly achievement during a circuits and electronics MOOC. In *Proceedings of the First ACM Conference on Learning@ Scale Conference* (pp. 169–170).

Dillon, J., Bosch, N., Chetlur, M., Wanigasekara, N., Ambrose, G. A., Sengupta, B., & D'Mello, S. K. (2016). Student emotion, co-occurrence, and dropout in a MOOC context. In *International Educational Data Mining Society*.

Dowell, N. M., Skrypnyk, O., Joksimovic, S., Graesser, A. C., Dawson, S., Gašević, D., Hennis, T. A., de Vries, P., & Kovanovic, V. (2015). Modeling learners' social centrality and performance through language and discourse. In *International Educational Data Mining Society*.

Fei, M., & Yeung, D.-Y. (2015). Temporal models for predicting student dropout in massive open online courses. In *2015 IEEE International Conference on Data Mining Workshop (ICDMW)* (pp. 256–263). IEEE.

Gardner, J., & Brooks, C. (2018). Student success prediction in MOOCs. *User Modeling and User-Adapted Interaction, 28*(2), 127–203.

Garman, G., et al. (2010). A logistic approach to predicting student success in online database courses. *American Journal of Business Education (AJBE), 3*(12), 1–6.

Greene, J. A., Oswald, C. A., & Pomerantz, J. (2015). Predictors of retention and achievement in a massive open online course. *American Educational Research Journal, 52*(5), 925–955.

He, J., Bailey, J., Rubinstein, B., & Zhang, R. (2015). Identifying at-risk students in massive open online courses. In *Proceedings of the AAAI Conference on Artificial Intelligence* (Vol. 29).

Jiang, S., Williams, A., Schenke, K., Warschauer, M., & O'dowd, D. (2014). Predicting MOOC performance with week 1 behavior. In *Educational Data Mining 2014*.

Joksimović, S., Manataki, A., Gašević, D., Dawson, S., Kovanović, V., & De Kereki, I. F. (2016). Translating network position into performance: Importance of centrality in different network configurations. In *Proceedings of the Sixth International Conference on Learning Analytics & Knowledge* (pp. 314–323).

Jordan, K. (2015). MOOC completion rates. *Recuperado de*. http://www.katyjordan.com/MOOCproject.html

Kennedy, G., Coffrin, C., De Barba, P., & Corrin, L. (2015). Predicting success: How learners' prior knowledge, skills and activities predict MOOC performance. In *Proceedings of the Fifth International Conference on Learning Analytics and Knowledge* (pp. 136–140).

Khalil, H., & Ebner, M. (2014). MOOCs completion rates and possible methods to improve retention-a literature review. In *EdMedia+ Innovate Learning* (pp. 1305–1313). Association for the Advancement of Computing in Education (AACE).

Kizilcec, R. F., & Halawa, S. (2015). Attrition and achievement gaps in online learning. In *Proceedings of the Second (2015) ACM Conference on Learning@ Scale* (pp. 57–66).

Kotsiantis, S., Tselios, N., Filippidi, A., & Komis, V. (2013). Using learning analytics to identify successful learners in a blended learning course. *International Journal of Technology Enhanced Learning, 5*(2), 133–150.
Mao, Y. (2018). Deep learning vs. Bayesian knowledge tracing: Student models for interventions. *Journal of Educational Data Mining, 10*(2).
Mwalumbwe, I., & Mtebe, J. S. (2017). Using learning analytics to predict students' performance in Moodle learning management system: A case of Mbeya University of Science and Technology. *The Electronic Journal of Information Systems in Developing Countries, 79*(1), 1–13.
Nagrecha, S., Dillon, J. Z., & Chawla, N. V. (2017). MOOC dropout prediction: Lessons learned from making pipelines interpretable. In *Proceedings of the 26th International Conference on World Wide Web Companion* (pp. 351–359).
Pardos, Z., Bergner, Y., Seaton, D., & Pritchard, D. (2013). Adapting Bayesian knowledge tracing to a massive open online course in edX. In *Educational Data Mining 2013*. Citeseer.
Pardos, Z. A., Tang, S., Davis, D., & Le, C. V. (2017). Enabling real-time adaptivity in MOOCs with a personalized next-step recommendation framework. In *Proceedings of the Fourth (2017) ACM Conference on Learning@ Scale* (pp. 23–32).
Pursel, B. K., Zhang, L., Jablokow, K. W., Choi, G. W., & Velegol, D. (2016). Understanding MOOC students: Motivations and behaviours indicative of MOOC completion. *Journal of Computer Assisted Learning, 32*(3), 202–217.
Qiu, J., Tang, J., Liu, T. X., Gong, J., Zhang, C., Zhang, Q., & Xue, Y. (2016). Modeling and predicting learning behavior in MOOCs. In *Proceedings of the Ninth ACM International Conference on Web Search and Data Mining* (pp. 93–102).
Ramos, C., & Yudko, E. (2008). "Hits" (not "discussion posts") predict student success in online courses: A double cross-validation study. *Computers & Education, 50*(4), 1174–1182.
Ren, Z., Rangwala, H., & Johri, A. (2016). Predicting performance on MOOC assessments using multi-regression models. Preprint, arXiv:1605.02269.
Rokach, L. (2005). *Data mining and knowledge discovery handbook*. Springer Science+ Business Media, Incorporated.
Rosé, C. P., Carlson, R., Yang, D., Wen, M., Resnick, L., Goldman, P., & Sherer, J. (2014). Social factors that contribute to attrition in MOOCs. In *Proceedings of the First ACM Conference on Learning@ Scale Conference* (pp. 197–198).
Russo, T. C., & Koesten, J. (2005). Prestige, centrality, and learning: A social network analysis of an online class. *Communication Education, 54*(3), 254–261.
Siemens, G., & Baker, R. S. d. (2012). Learning analytics and educational data mining: Towards communication and collaboration. In *Proceedings of the 2nd International Conference on Learning Analytics and Knowledge* (pp. 252–254).
Sinha, T., Li, N., Jermann, P., & Dillenbourg, P. (2014). Capturing" attrition intensifying" structural traits from didactic interaction sequences of MOOC learners. Preprint, arXiv:1409.5887.
Stein, R. M., & Allione, G. (2014). Mass attrition: An analysis of drop out from a principles of microeconomics MOOC.
Tucker, C., Pursel, B. K., & Divinsky, A. (2014). Mining student-generated textual data in MOOCs and quantifying their effects on student performance and learning outcomes. In *2014 ASEE Annual Conference & Exposition* (pp. 24–907).
UNESCO. (2005). Rapport mondial de suivi sur l'ept 2005. *UNESCO* (4), 57–84.
UNESCO, Mundial, G. B., UNICEF, et al. (2016). Education 2030: Incheon declaration and framework for action: Towards inclusive and equitable quality education and lifelong learning for all.
Wang, X., Yang, D., Wen, M., Koedinger, K., & Rosé, C. P. (2015). Investigating how student's cognitive behavior in MOOC discussion forums affect learning gains. In *International Educational Data Mining Society*.
Wen, M., Yang, D., & Rose, C. (2014). Sentiment analysis in MOOC discussion forums: What does it tell us? In *Educational Data Mining 2014*.

Whitehill, J., Williams, J., Lopez, G., Coleman, C., & Reich, J. (2015). Beyond prediction: First steps toward automatic intervention in MOOC student stopout. Available at SSRN 2611750.
Wolpert, D. H., & Macready, W. G. (1997). No free lunch theorems for optimization. *IEEE Transactions on Evolutionary Computation, 1*(1), 67–82.
Xiao, X., Pham, P., & Wang, J. (2015). Attentivelearner: Adaptive mobile MOOC learning via implicit cognitive states inference. In *Proceedings of the 2015 ACM on International Conference on Multimodal Interaction* (pp. 373–374).
Xing, W., Chen, X., Stein, J., & Marcinkowski, M. (2016). Temporal predication of dropouts in MOOCs: Reaching the low hanging fruit through stacking generalization. *Computers in Human Behavior, 58*, 119–129.
Xing, W., & Du, D. (2019). Dropout prediction in MOOCs: Using deep learning for personalized intervention. *Journal of Educational Computing Research, 57*(3), 547–570.

Chapter 4
Rainfall Prediction Using Machine Learning Models: Literature Survey

Eslam A. Hussein, Mehrdad Ghaziasgar, Christopher Thron, Mattia Vaccari, and Yahlieel Jafta

4.1 Introduction

Natural processes on Earth can be classified into several categories, including hydrological processes like storm waves and groundwater; biological processes like forest growth; atmospheric processes like thunderstorms and rainfall; human processes like urban development; and geological processes like earthquakes. The field of physical geography seeks to investigate the distribution of the different features/parameters that describe the landscape and functioning of the Earth by analyzing the processes that shape it. These features/parameters have been referred to as geophysical parameters in the literature (Karimi, 2014).

Rainfall is a key geophysical parameter that is essential for many applications in water resource management, especially in the agriculture sector. Predicting rainfall can help managers in various sectors to make decisions regarding a range of important activities such as crop planting, traffic control, the operation of sewer systems, and managing disasters like droughts and floods (Htike and Khalifa, 2010). A number of countries such as Malaysia and India depend on the agriculture sector as a major contributor to the economy (Htike and Khalifa, 2010; Parmar et al., 2017) and as a source of food security. Hence, an accurate prediction of rainfall is needed

E. A. Hussein (✉) · M. Ghaziasgar · Y. Jafta
Department of Computer Science, University of the Western Cape, Cape Town, South Africa
e-mail: ehussein@uwc.ac.za; mghaziasgar@uwc.ac.za; 2858132@myuwc.ac.za

C. Thron
Department of Science and Mathematics, Texas A&M University-Central, Killeen, TX, USA
e-mail: thron@tamuct.edu

M. Vaccari
Department of Physics and Astronomy, University of the Western Cape, Cape Town, South Africa
e-mail: mvaccari@uwc.ac.za

M. Alloghani et al. (eds.), *Artificial Intelligence for Data Science in Theory and Practice*, Studies in Computational Intelligence 1006,
https://doi.org/10.1007/978-3-030-92245-0_4

to make better future decisions to help manage activities such as the ones mentioned above.

Rainfall is considered to be one of the most complicated parameters to forecast in the hydrological cycle (Htike and Khalifa, 2010; Hung et al., 2009; Nasseri et al., 2008). This is due to the dynamic nature of environmental factors and random variations, both spatially and temporally, in these factors (Htike and Khalifa, 2010). Therefore, to address random variations in rainfall, several machine learning (ML) tools including artificial neural networks (ANN), *k*-nearest neighbours (KNNs), decision trees (DT), etc. are used in the literature to learn patterns in the data to forecast rainfall. In this chapter, a review of past work in the area of rainfall prediction using ML models is carried out.

A number of related review papers exist as follows. The authors in Mosavi et al. (2018) focused on reviewing studies that use ML for flood prediction, which closely resembles rainfall prediction. The authors in Shi and Yeung (2018) focused on the use of ML for generic spatio-temporal sequence forecasting. Finally, the authors in Parmar et al. (2017) conducted a survey on the use of ML for rainfall prediction: however, the study was limited to rainfall prediction in India.

This chapter serves as an addition to the field by surveying recent relevant studies focusing on the use of ML in rainfall prediction in a variety of geographic locations from 2016–2020. After detailing the methods used to forecast rainfall, one of the important contributions of this chapter is to demonstrate various pitfalls that lead to an overestimation in model performance of the ML models in various papers. This in turn leads to unrealistic hype and expectations surrounding ML in the current literature. It also leads to an unrealistic understanding of the advancements in, and gains by, ML research in this field. It is therefore important to clearly state and demonstrate these pitfalls in order to help researchers avoid them.

The rest of this review is organized as follows: Section 4.2 discusses the methodology used to survey and review the literature which defines the discussion framework used in all subsequent sections; Sect. 4.3 describes the data sets used; Sect. 4.4 provides a description of the output objective in the various papers; Sects. 4.5–4.7 describe the input features used, common methods of pre-processing and the ML models used; Sect. 4.8 summarizes the results obtained in various studies; and Sect. 4.9 then provides a discussion of the procedures used, specifically pointing out the pitfalls mentioned before towards obtaining over-estimated and unrealistic results. The section that follows concludes the paper.

4.2 Methodology

This chapter carries out an in-depth review of relevant literature to reveal the different practices authors take to predict rainfall. The review covers several aspects which relate to the input into, output from, and methods used in the various systems devised in the literature for this purpose. The review specifically focuses on studies that use supervised learning for both regression and classification problems.

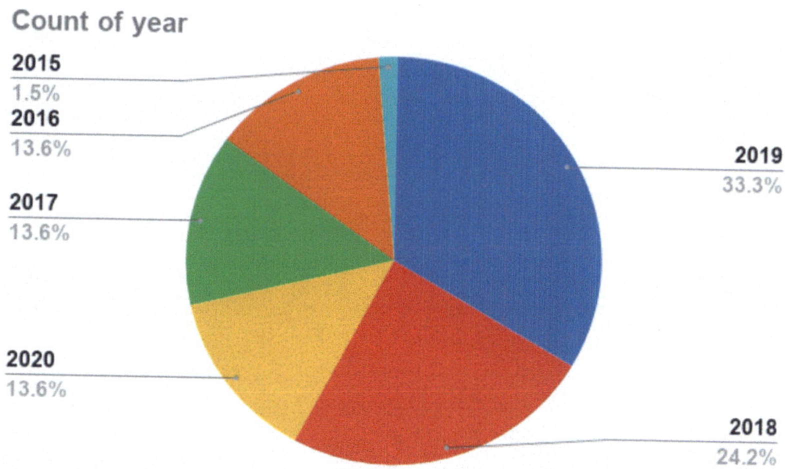

Fig. 4.1 Pie chart showing proportions by publication year for papers in this review

Google scholar was used to collect papers from 2016 to 2020, with the following key words: ("machine learning" OR "deep learning") AND ("precipitation prediction" OR "rainfall prediction" OR "precipitation nowcasting"). Almost 1240 results were obtained, and of these only supervised rainfall prediction papers that used meteorological data from, e.g., radar, satellites, and stations were selected, while papers that used data from normal cameras, e.g., photographs were excluded. Even though this review focuses on the prediction of rainfall, the methods used to achieve this can be extended and applied to other geophysical parameters like temperature and wind. Hence, the conclusions and discussions of this chapter can be adapted to other parameters.

The total number of reviewed papers are 66, which are a combination of conferences and journal papers published from 2016–2020, except for one paper (Shi et al., 2015) which was published in 2015 and is a seminal work in this field. Figure 4.1 shows the reviewed studies per year. Tables which summaries the reviewed paper can be found in Appendices 1 and 2.

Figure 4.2 shows the generic structure of supervised ML models. This structure was used as a guideline to construct a set of questions used to systematically categorize and analyze the 66 papers. The questions are as follows:

1. What data sets are used and where are they sourced?
2. What is the output objective in the various papers in terms of what the goal of prediction/forecasting?
3. What input features are extracted from the data set(s) to be used to achieve the output objective?
4. What pre-processing methods are used prior to classification/regression?
5. What ML models are used to achieve classification/regression towards the output objective?

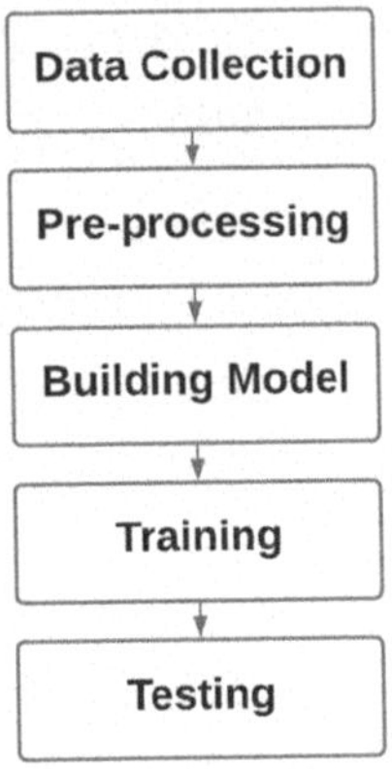

Fig. 4.2 Basic flow for building machine learning (ML) models (Mosavi et al., 2018)

6. What results were obtained from the above-mentioned steps, and how were they reported?

These questions provide the framework for the rest of this paper. Sections 4.3–4.8 address questions 1–6 in sequence. Section 4.9 discusses the findings in the previous six sections, and Sect. 4.10 provides conclusions.

4.3 Data Sets

This section provides a breakdown of the data sets used in the 66 studies surveyed, based on the sources of the data sets, availability, and geographical locations where the data sets were collected.

Figure 4.3 (left) provides a breakdown of the studies based on the sources/availability of the data sets used in those studies. About 75% of the studies used private data, sourced from meteorological stations of their prospective countries (Peng et al., 2019; Pham et al., 2019; Zhang et al., 2020, 2018; Zainudin et al., 2016; Singh and Kumar, 2019; Oswal, 2019; Balamurugan and Manojkumar, 2021; Du et al., 2017; Manandhar et al., 2019; Kashiwao et al., 2017; Shi et al., 2015; Singh et al., 2017; Jing et al., 2019; Ayzelet al., 2019; Chen et al., 2020; Tran and Song, 2019a; Shi et al., 2017; Wang et al., 2017; Tran and Song, 2019b; Shi et al., 2017; Du et al., 2018; Chattopadhyay et al., 2020; Zhuang and Ding, 2016; Du et al., 2019; Dash et al. , 2018; Cristian, 2018; Lakshmaiah et al., 2016; Ramsundram et al., 2016; Sulaiman and Wahab, 2018; Amiri et al., 2016; Abbot and Marohasy, 2016; Sardeshpande and Thool, 2019; Shenify et al., 2016; Banadkooki et al. , 2019; Mehr et al., 2019; Beheshti et al., 2016; Nourani et al., 2019; Abbot and Marohasy, 2017; Kumar et al., 2019; Haidar and Verma, 2018; Duong et al., 2018; Xu et al., 2020; Aswin et al., 2018; Chhetri et al., 2020; Bojang et al., 2020; Canchala et al., 2020; Mehdizadeh et al., 2018; Weesakul et al., 2018; Gao et al., 2019; Mishra and Kushwaha, 2019). Most of these data sets are not readily available for use. Only

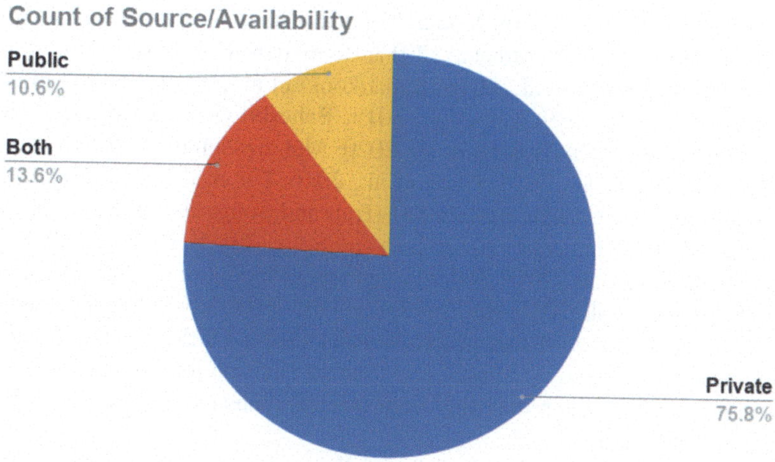

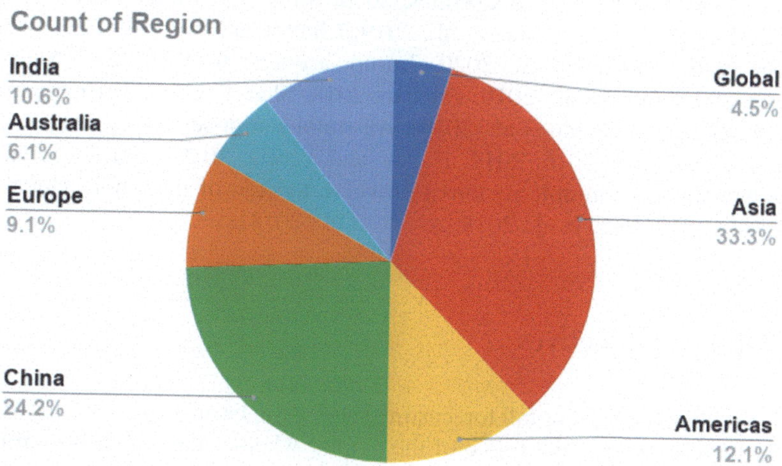

Fig. 4.3 Pie chart of the percentage of data sets in this survey in terms of source/availability (top) and geographical region (bottom)

10% of the studies use data sourced from freely available sources such as Kaggle (www.kaggle.com), and the National Oceanic and Atmospheric Administration (NOAA) (Sato et al., 2018; Castro et al., 2020; Pan et al., 2019; Zhan et al., 2019; Patel et al., 2018; Damavandi and Shah, 2019; Aswin et al., 2018). The remaining 13% of studies in this review use data from both private and publicly available sources (Valencia-Payan and Corrales, 2018; Yu et al., 2017; Diez-Sierra and del Jesus, 2020; Chen et al., 2016; Huang et al., 2017; Boonyuen et al., 2018, 2019; Lee et al., 2018; Aguasca-Colomo et al., 2019).

Figure 4.3 (right) summarizes the geographical regions included in this review. The continent of Asia accounts for around 68% of all studies (Pham et al., 2019; Yu et al., 2017; Zainudin et al., 2016; Manandhar et al., 2019; Kashiwao et al., 2017; Shi et al., 2015; Sato et al., 2018; Shi et al., 2017; Boonyuen et al., 2018, 2019; Sulaiman and Wahab, 2018; Amiri et al., 2016; Banadkooki et al. , 2019; Mehr et al., 2019; Damavandi and Shah, 2019; Lee et al., 2018; Beheshti et al., 2016; Duong et al., 2018; Chhetri et al., 2020; Bojang et al., 2020; Mehdizadeh et al., 2018; Weesakul et al., 2018; Zhang et al., 2018; Chen et al., 2016; Du et al., 2017; Huang et al., 2017; Jing et al., 2019; Chen et al., 2020; Tran and Song, 2019a; Shi et al., 2017; Castro et al., 2020; Wang et al., 2017; Tran and Song, 2019b; Du et al., 2018, 2019; Xu et al., 2020; Gao et al., 2019; Balamurugan and Manojkumar, 2021; Dash et al. , 2018; Lakshmaiah et al., 2016; Ramsundram et al., 2016; Sardeshpande and Thool, 2019; Kumar et al., 2019; Mishra and Kushwaha, 2019). Of these, studies that focus on China and India make up almost one quarter and one tenth respectively of all studies in this review. The remaining Asian studies focus on countries such as Iran, South Korea and Japan.

The rest of the chart is distributed as follows: the Americas make up 12.1% of studies (Valencia-Payan and Corrales, 2018; Singh and Kumar, 2019; Singh et al., 2017; Pan et al., 2019; Zhan et al., 2019; Chattopadhyay et al., 2020; Zhuang and Ding, 2016; Canchala et al., 2020); Europe accounts for 9.1% (Diez-Sierra and del Jesus, 2020; Ayzelet al., 2019; Cristian, 2018; Shenify et al., 2016; Nourani et al., 2019; Aguasca-Colomo et al., 2019); Australia comprises 6.1%; (Oswal, 2019; Abbot and Marohasy, 2016, 2017; Haidar and Verma, 2018), and the remaining 4.5% either involve multiple regions or involve the use of the whole global map (Peng et al., 2019; Patel et al., 2018; Aswin et al., 2018).

4.4 Output Objectives

The output objectives of rainfall forecasting studies can be analyzed in terms of three factors: the forecasting time frame of the output; whether the output is continuous or discrete; and the dimensionality of the output. The forecasting time frame of the output specifies the time span of the forecast made, i.e., hourly, daily, monthly, etc. The output can also be discrete (e.g., classification into "Rain"/"No Rain" classes), or continuous (e.g., predicting the quantity of rain), or both. Finally, the output can be 1-dimensional (1D) in the form of a single number or label representing a rainfall measure or category, or 2-dimensional (2D) in the form of a geospatial map of rainfall measures or categories on a grid of the geographical location under study.

In terms of the forecasting time frame, the studies can be broken down into those that make long-term predictions and those that focus on making short-term predictions. In this review, long-term prediction is defined as predictions of one month up to a year ahead, while short-term prediction can be a few minutes ahead (e.g., 5–15 minutes), up to one or more days ahead. Figure 4.4 (left) shows the distribution of papers' forecasting time frames. Of the 66 reviewed papers, 30 papers (45%) make long-term predictions, the majority of which focus on

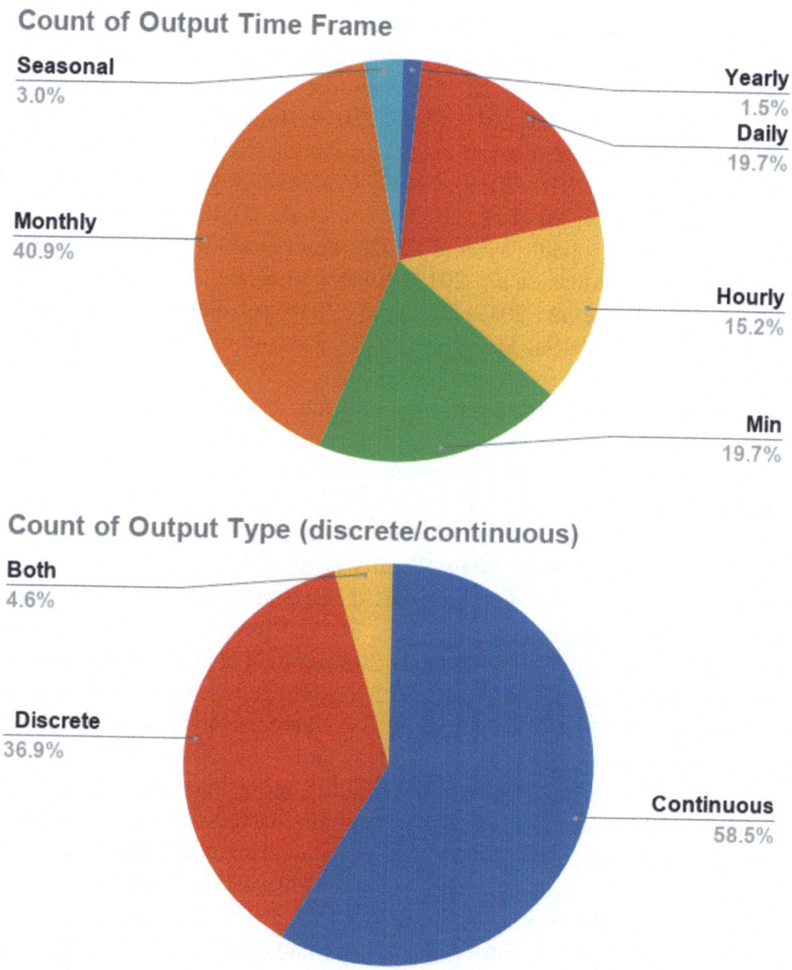

Fig. 4.4 Pie chart of the percentage of data sets in this survey in terms of forecasting time frame (top) and the discrete (classification)/continuous (regression) nature of the prediction output (bottom)

monthly forecasting (Cristian, 2018; Lakshmaiah et al., 2016; Ramsundram et al., 2016; Sulaiman and Wahab, 2018; Amiri et al., 2016; Abbot and Marohasy, 2016; Sardeshpande and Thool, 2019; Shenify et al., 2016; Banadkooki et al. , 2019; Mehr et al., 2019; Damavandi and Shah, 2019; Lee et al., 2018; Beheshti et al., 2016; Nourani et al., 2019; Abbot and Marohasy, 2017; Kumar et al., 2019; Haidar and Verma, 2018; Duong et al., 2018; Xu et al., 2020; Aswin et al., 2018; Canchala et al., 2020; Bojang et al., 2020; Chhetri et al., 2020; Mehdizadeh et al., 2018; Weesakul et al., 2018; Mishra and Kushwaha, 2019; Aguasca-Colomo et al., 2019). Only two studies focus on seasonal forecasting (Du et al., 2019; Dash et al. , 2018),while a single study aims towards yearly forecasting (Gao et al., 2019). As for studies that focus on short-term prediction, these are broken down nearly evenly between daily (Peng et al., 2019; Pham et al., 2019; Zainudin et al., 2016; Diez-Sierra and del Jesus, 2020; Singh and Kumar, 2019; Oswal, 2019; Balamurugan and Manojkumar, 2021; Huang et al., 2017; Castro et al., 2020; Pan et al., 2019; Zhuang and Ding, 2016; Boonyuen et al., 2018, 2019), hourly (Valencia-Payan and Corrales, 2018; Zhang et al., 2020, 2018; Yu et al., 2017; Chen et al., 2016; Du et al., 2017, 2018; Zhan et al., 2019; Chattopadhyay et al., 2020; Patel et al., 2018), and one or more minutes ahead (Manandhar et al., 2019; Kashiwao et al., 2017; Shi et al., 2015; Singh et al., 2017; Jing et al., 2019; Sato et al., 2018; Ayzelet al., 2019; Chen et al., 2020; Tran and Song, 2019a; Shi et al., 2017; Wang et al., 2017; Tran and Song, 2019b; Shi et al., 2017).

In terms of the type of output, i.e., discrete (classification) or continuous (regression), Fig. 4.4 (right) shows the distribution between the different output types. The majority carried out regressions to obtain continuous output (Du et al., 2019; Dash et al. , 2018; Cristian, 2018; Lakshmaiah et al., 2016; Ramsundram et al., 2016; Sulaiman and Wahab, 2018; Amiri et al., 2016; Abbot and Marohasy, 2016; Sardeshpande and Thool, 2019; Shenify et al., 2016; Banadkooki et al. , 2019; Mehr et al., 2019; Damavandi and Shah, 2019; Lee et al., 2018; Beheshti et al., 2016; Nourani et al., 2019; Abbot and Marohasy, 2017; Kumar et al., 2019; Haidar and Verma, 2018; Duong et al., 2018; Xu et al., 2020; Aswin et al., 2018; Canchala et al., 2020; Bojang et al., 2020; Chhetri et al., 2020; Mehdizadeh et al., 2018; Weesakul et al., 2018; Peng et al., 2019; Pham et al., 2019; Valencia-Payan and Corrales, 2018; Zhang et al., 2020, 2018; Yu et al., 2017; Tran and Song, 2019a; Wang et al., 2017; Shi et al., 2017; Du et al., 2018), while slightly more than one third carried out classification into discrete classes (Zainudin et al., 2016; Diez-Sierra and del Jesus, 2020; Singh and Kumar, 2019; Oswal, 2019; Balamurugan and Manojkumar, 2021; Chen et al., 2016; Du et al., 2017; Manandhar et al., 2019; Kashiwao et al., 2017; Huang et al., 2017; Shi et al., 2015; Singh et al., 2017; Jing et al., 2019; Sato et al., 2018; Chen et al., 2020; Shi et al., 2017; Chattopadhyay et al., 2020; Patel et al., 2018; Zhuang and Ding, 2016; Boonyuen et al., 2018, 2019; Gao et al., 2019; Mishra and Kushwaha, 2019; Aguasca-Colomo et al., 2019). Only 3 studies applied both classification and regression (Ayzelet al., 2019; Tran and Song, 2019b; Zhan et al., 2019).

For studies that applied classification,mostly carried out binary classification (Zainudin et al., 2016; Diez-Sierra and del Jesus, 2020; Singh and Kumar, 2019; Oswal, 2019; Balamurugan and Manojkumar, 2021; Chen et al., 2016; Du et al., 2017; Manandhar et al., 2019; Kashiwao et al., 2017; Huang et al., 2017; Shi et al., 2015; Singh et al., 2017; Jing et al., 2019; Sato et al., 2018; Chen et al., 2020; Shi et al., 2017; Patel et al., 2018; Zhuang and Ding, 2016; Boonyuen et al., 2018), with the majority of these classified into "Rain"/"No Rain" classes. Relatively fewer studies aim towards carrying out classification into multiple classes (Huang et al., 2017; Chattopadhyay et al., 2020; Boonyuen et al., 2019; Gao et al., 2019; Mishra and Kushwaha, 2019), varying from three to five classes.

Finally, for the dimensionality of the output, 54 out of 66 studies produce 1D output (Du et al., 2019; Dash et al. , 2018; Banadkooki et al. , 2019; Mehr et al., 2019; Damavandi and Shah, 2019; Lee et al., 2018; Beheshti et al., 2016; Nourani et al., 2019; Abbot and Marohasy, 2017; Kumar et al., 2019; Haidar and Verma, 2018; Gao et al., 2019; Mishra and Kushwaha, 2019; Zhang et al., 2020, 2018; Yu et al., 2017; Zainudin et al., 2016; Diez-Sierra and del Jesus, 2020; Singh and Kumar, 2019; Oswal, 2019; Aguasca-Colomo et al., 2019; Peng et al., 2019; Pham et al., 2019; Valencia-Payan and Corrales, 2018; Balamurugan and Manojkumar, 2021; Chen et al., 2016; Du et al., 2017; Manandhar et al., 2019; Kashiwao et al., 2017; Huang et al., 2017; Pan et al., 2019; Du et al., 2018; Zhan et al., 2019; Chattopadhyay et al., 2020; Patel et al., 2018; Zhuang and Ding, 2016; Boonyuen et al., 2018, 2019), with the remaining 12 studies producing a series 2D images as output (Shi et al., 2015; Singh et al., 2017; Jing et al., 2019; Sato et al., 2018; Ayzelet al., 2019; Chen et al., 2020; Tran and Song, 2019a; Shi et al., 2017; Castro et al., 2020; Wang et al., 2017; Tran and Song, 2019b; Shi et al., 2017). Of the studies with 2D output, all except one (Castro et al., 2020) involve short-term prediction intervals of 10 minutes or less.

A connection between forecasting time frame and the discrete/continuous nature of the output can be observed. In general, studies involving longer-term predictions tend to make use of regression which produces continuous output, whereas on short-term time frames, studies tend towards using classification that gives discrete output. Specifically, 27 out of the 30 papers that focus on long-term prediction carry out regression (Du et al., 2019; Dash et al. , 2018; Cristian, 2018; Lakshmaiah et al., 2016; Ramsundram et al., 2016; Sulaiman and Wahab, 2018; Amiri et al., 2016; Abbot and Marohasy, 2016; Sardeshpande and Thool, 2019; Shenify et al., 2016; Banadkooki et al. , 2019; Mehr et al., 2019; Damavandi and Shah, 2019; Lee et al., 2018; Beheshti et al., 2016; Nourani et al., 2019; Abbot and Marohasy, 2017; Kumar et al., 2019; Haidar and Verma, 2018; Duong et al., 2018; Xu et al., 2020; Aswin et al., 2018; Canchala et al., 2020; Bojang et al., 2020; Chhetri et al., 2020; Mehdizadeh et al., 2018; Weesakul et al., 2018), and 23 of the 36 papers that focus on short-term prediction carry out classification (Zainudin et al., 2016; Diez-Sierra and del Jesus, 2020; Singh and Kumar, 2019; Oswal, 2019; Balamurugan and Manojkumar, 2021; Chen et al., 2016; Du et al., 2017; Manandhar et al., 2019; Kashiwao et al., 2017; Huang et al., 2017; Shi et al., 2015; Singh et al., 2017; Jing et al., 2019; Sato et al., 2018; Ayzelet al., 2019; Chen et al., 2020; Shi et al., 2017; Zhan et al., 2019; Chattopadhyay et al., 2020; Patel et al., 2018; Zhuang and Ding, 2016; Boonyuen et al., 2018, 2019). This relation may be explained by the fact that longer-term studies usually aim at predicting averages over several days (up to a month), while short-term studies predict instantaneous conditions. Multi-day averaged data assumes a continuous range of values, while in instantaneous rainfall data sets most values are null. It follows that classification into rain/no rain is useful for short term, but not for long-term prediction.

4.5 Input Features

In order to make future predictions, studies make use of data from one or more time steps (called "lags" or "time lags") as input features to predict one or more future lags. For example, to predict rainfall at lag T, two previous time lags $(T - 1)$ and $(T - 2)$ may be used.

The actual input features in each lag vary across studies. In general, the input features used in the studies in this review were found to be of two types: 1D input features in which each time lag in the data set represents one or a set of geophysical parameters that have been collected at static known locations, i.e., meteorological stations; and 2D input features in which each time lag in the data set is a 2D spatial map of values representing rainfall in the geographical area under review, usually collected by satellite or radar.

1D input features used include geophysical parameters such as temperature, humidity, wind speed and air pressure (Ramsundram et al., 2016; Amiri et al., 2016; Banadkooki et al. , 2019; Damavandi and Shah, 2019; Aguasca-Colomo et al., 2019; Oswal, 2019; Manandhar et al., 2019; Kashiwao et al., 2017; Lu et al., 2018; Xu et al., 2020). In a smaller number of cases, climatic indices such as the Pacific Decadal Oscillation may also be used (Du et al., 2019; Abbot and Marohasy, 2016; Lee et al., 2018; Gao et al., 2019; Valencia-Payan and Corrales, 2018). Studies that use 1D input features tend to use a relatively small number of overall input features, ranging from 2–12 features used for prediction.

With 2D input features, one or more images are taken as input features, depending on the number of time lags used as input, e.g., two time lags used as input implies that two images are used as input. The number of time lags used as input is henceforth referred to as the "sequence length."

There is no rule of thumb for how many time lags should be used as input, and this is mostly selected arbitrarily, and in fewer cases via trial and error. The vast majority of the studies under review select a fixed sequence length. The sequence length can be viewed as a hyper-parameter that affects the prediction outcome, but the optimization of this hyper-parameter is not investigated in the studies under review. The studies under review were found to be more focused on the machine learning component, mostly at devising new deep learning architectures, than selecting and tuning other aspects of their systems.

The most common sequence lengths used are 5 frames (Singh et al., 2017; Jing et al., 2019; Sato et al., 2018; Wang et al., 2017) and 10 frames (Singh et al., 2017; Jing et al., 2019; Sato et al., 2018; Wang et al., 2017). Other sequence lengths are also used, such as 2 (Singh et al., 2017), 4 (Sato et al., 2018), 7 (Wang et al., 2017) and 20 (Jing et al., 2019).

Studies that use 2D input features tend to use a relatively large number of input features. This can be attributed to the fact that the feature vectors produced are associated with one or more 2D images, resulting in vectors of size (Image width × Image height × Sequence length). Overall, the number of features can grow as high as several thousands.

Typically, 1D or 2D inputs are used to predict 1D or 2D outputs, respectively. As noted in the previous section, longer-term predictions tend to make 1D predictions, so it follows these studies also tend to use 1D data (Du et al., 2019; Dash et al. , 2018; Cristian, 2018; Lakshmaiah et al., 2016; Ramsundram et al., 2016; Sulaiman and Wahab, 2018; Amiri et al., 2016; Abbot and Marohasy, 2016; Sardeshpande and Thool, 2019; Banadkooki et al. , 2019; Mehr et al., 2019; Damavandi and Shah, 2019; Lee et al., 2018; Beheshti et al., 2016; Nourani et al., 2019; Abbot and Marohasy, 2017; Kumar et al., 2019; Duong et al., 2018; Xu et al., 2020; Canchala et al., 2020; Bojang et al., 2020; Chhetri et al., 2020; Mehdizadeh et al., 2018; Mishra and Kushwaha, 2019; Aguasca-Colomo et al., 2019), while those that make shorter-term predictions tend towards the use of 2D data (Shi et al., 2015; Singh et al., 2017; Jing et al., 2019; Sato et al., 2018; Ayzelet al., 2019; Chen et al., 2020; Tran and Song, 2019a; Shi et al., 2017; Castro et al., 2020; Wang et al., 2017; Tran and Song, 2019b; Shi et al., 2017; Boonyuen et al., 2019)

4.6 Input Data Pre-processing

Before ML tools are applied to make predictions on the available data, the input data is usually pre-processed to reformat the data into a form that will make training of, and prediction by, the ML tool(s) easier and faster. The pre-processing techniques usually applied in geophysical parameter forecasting can be broken down into three broad categories, namely data imputation; feature selection/reduction; and data preparation for classification. The following subsections describe these categories, as well as their application in the papers in this review.

4.6.1 Data Imputation

Data sets are regularly found to have missing data entries, which is caused by a range of factors such as data corruption, data sensor malfunction, etc. This is a serious issue faced by researchers in data mining or analysis and needs to be addressed as part of pre-processing before feature selection/preparation and training.

The techniques used to infer and substitute missing data are collectively referred to as data imputation techniques. Data imputation is challenging and is an on-going research area. In the papers in this review, it was found that very little focus was placed on this problem, with most of the studies making use of simple statistical techniques such as averaging to interpolate missing data entries (Sulaiman and Wahab, 2018; Haidar and Verma, 2018; Canchala et al., 2020; Bojang et al., 2020; Zainudin et al., 2016; Oswal, 2019). While not used in the papers in this review, more advanced data imputation techniques exist beyond the use of simple statistics, such as the use of ML to impute the data. The interested reader may refer to (Tang and Ishwaran, 2017; Shah et al., 2014; Pantanowitz and Marwala, 2009).

4.6.2 *Feature Selection/Reduction*

Feature selection/reduction aims to determine and use salient features in the data, and disregard irrelevant features in the data. This helps to reduce training time, decrease the model complexity and increase its performance. In the papers in this review, it is observed that feature selection is carried out either automatically or manually.

For automatic feature selection, various algorithms are used to determine the most salient features in the data. The most common method used in the papers in this review involved the use of deep learning techniques such as ANNs and convolutional neural networks (CNNs), to select/reduce features automatically, most especially when high-dimensional data such as radar and satellite images was used (Shi et al., 2015; Singh et al., 2017; Jing et al., 2019; Sato et al., 2018; Chen et al., 2020; Shi et al., 2017; Castro et al., 2020; Wang et al., 2017; Tran and Song, 2019b; Shi et al., 2017; Patel et al., 2018; Zhuang and Ding, 2016; Boonyuen et al., 2018, 2019; Haidar and Verma, 2018; Weesakul et al., 2018). The use of deep learning techniques was found to be much more common with short-term data sets which are generally much larger, therefore making it possible to achieve convergence on deep networks. Another category of ML tools used for automatic feature selection includes ensemble methods like random forests (RFs) which automatically order features in terms of importance, as used in Damavandi and Shah (2019), Bojang et al. (2020); Aguasca-Colomo et al. (2019), Valencia-Payan and Corrales (2018), Yu et al. (2017), Zainudin et al. (2016), Diez-Sierra and del Jesus (2020), Singh and Kumar (2019), and Balamurugan and Manojkumar (2021). Finally, principle components analysis (PCA) has also been used to reduce features in Du et al. (2019), Peng et al. (2019), Zhang et al. (2018), Diez-Sierra and del Jesus (2020), Du et al. (2017), and Pan et al. (2019).

As regards manual feature selection, researchers may either use prior experience and trial and error to manually select relevant features such as in Dash et al. (2018), Cristian (2018), Dash et al. (2018), Cristian (2018), Lakshmaiah et al. (2016), Sulaiman and Wahab (2018), Sardeshpande and Thool (2019), Shenify et al. (2016), Banadkooki et al. (2019), Nourani et al. (2019), Haidar and Verma (2018), Duong et al. (2018), and Canchala et al. (2020) or use correlation analysis methods such as auto correlation to indirectly inform the manual feature selection process as in Du et al. (2019), Ramsundram et al. (2016), Abbot and Marohasy (2016), Mehr et al. (2019), Damavandi and Shah (2019), Lee et al. (2018), Abbot and Marohasy (2017), Kumar et al. (2019), and Gao et al. (2019). Where images are used, image cropping and resizing is applied to, respectively, dispose of irrelevant/static image segments and reduce the number of features (Shi et al., 2015; Singh et al., 2017; Jing et al., 2019; Sato et al., 2018; Chen et al., 2020; Shi et al., 2017; Castro et al., 2020; Wang et al., 2017; Tran and Song, 2019b; Shi et al., 2017).

Manual feature selection is much more common with long-term data sets, with very few long-term prediction studies in this review making use of automatic feature selection methods. This is partly attributed to the relatively smaller amount of data available in these sets, as mentioned before, which makes it challenging, or even rules out, the application of, e.g., deep learning methods for automatic feature selection.

The rotation of the earth around the sun can cause data to exhibit a seasonal behavior on an annual basis, i.e., they exhibit annual periodicity (Delleur and Kavvas, 1978; Barnettet al., 2012). This is most prominent in long-term data sets and less prominent in shorter-term data sets. Addressing seasonality in long-term data sets is critical when traditional time series models are used, since these models assume stationarity (Delleur and Kavvas, 1978; Nielsen, 2020), while seasonality and trends in general makes time series non-stationary. Converting data from a non-stationary to a stationary state involves is a process of generating a time series with statistical properties that do not change over time. For further information about seasonal and non-stationary data sets and the conversion of non-stationary to stationary time series, the interested reader is referred to (Nielsen, 2020). Another way to deal with seasonality is the inclusion of features that exhibit seasonal behavior, such as the usage of the same month previous year.

Figure 4.5 shows the methodologies used in the long-term prediction studies in this review. 11 of the 30 long-term papers (37%) did not address seasonality in the data (Du et al., 2019; Dash et al. , 2018; Ramsundram et al., 2016; Sardeshpande and Thool, 2019; Banadkooki et al. , 2019; Damavandi and Shah, 2019; Lee et al., 2018; Mehdizadeh et al., 2018; Mishra and Kushwaha, 2019; Aguasca-Colomo et al., 2019; Gao et al., 2019),while the remaining 19 papers used some means of addressing seasonality in the data (Bojang et al., 2020; Amiri et al., 2016; Shenify et al., 2016; Xu et al., 2020; Mehr et al., 2019; Beheshti et al., 2016; Canchala et al., 2020; Cristian, 2018; Lakshmaiah et al., 2016; Sulaiman and Wahab, 2018; Abbot and Marohasy, 2016; Nourani et al., 2019; Abbot and Marohasy, 2017; Kumar et al., 2019; Haidar and Verma, 2018; Duong et al., 2018; Aswin et al., 2018; Chhetri et al., 2020; Weesakul et al., 2018).

In the papers that addressed seasonality, four unique approaches were identified, and some were combined with others. The first approach involves including features from lag $T-12$ (same month previous year) in the feature set used to predict rainfall at month T (Mehr et al., 2019; Beheshti et al., 2016; Canchala et al., 2020; Cristian, 2018; Lakshmaiah et al., 2016; Sulaiman and Wahab, 2018; Abbot and Marohasy, 2016; Nourani et al., 2019; Abbot and Marohasy, 2017; Kumar et al., 2019; Haidar and Verma, 2018; Duong et al., 2018; Aswin et al., 2018; Chhetri et al., 2020; Weesakul et al., 2018). A less common approach is to use the index of the current month in the year (1=January, ..., 12=December) as an input feature (Haidar and Verma, 2018; Beheshti et al., 2016).

Alternative approaches include performing time series decomposition, either using singular spectrum analysis as in Bojang et al. (2020) or wavelet transformation as in Amiri et al. (2016), Xu et al. (2020), and Shenify et al. (2016). One paper (Beheshti et al., 2016) combined time series decomposition using singular spectrum analysis with the inclusion of features from lag $T-12$ in the feature set. This has been included in the segment labelled "Combination" in Fig. 4.5.

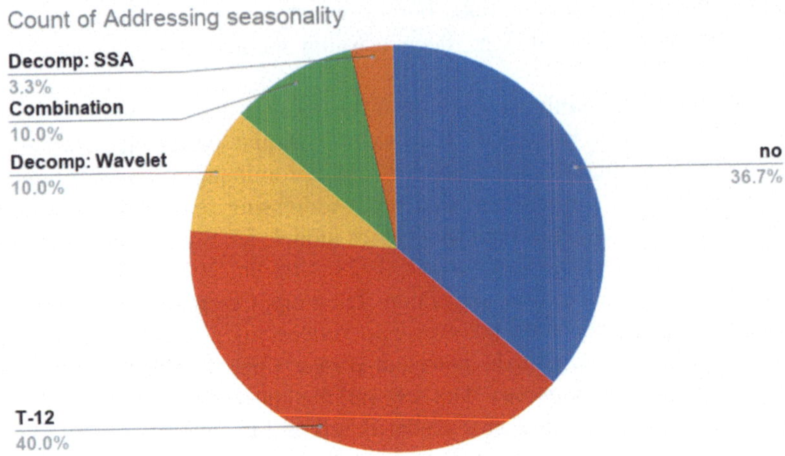

Fig. 4.5 Methods used to account for seasonality in studies with long-term data, by percentage

The final approach used to address seasonality takes the form of data de-seasonalization by subtracting the monthly averages from the data as in Chhetri et al. (2020) and Mehr et al. (2019). All of the papers in this review that used this approach combined this subtraction with the first approach, i.e., including features from lag $T - 12$ in the feature set. These two papers have also been included in the segment labelled "Combination" in Fig. 4.5.

4.6.3 Data Preparation for Classification

When attempting to carry out classification into discrete classes, it is either necessary to use a data set in which the desired output variable is discrete or to convert a desired continuous-valued output variable into discrete classes. This involves setting the desired number of classes, which is usually done manually and arbitrarily, followed by determining the range of values represented by each class, i.e., determining the thresholds that divide the continuous scale into the desired classes. Finally, where the number of instances across classes is imbalanced, it is necessary to balance them.

In the papers in this survey that carried out classification, most made use of data that was continuous, yet very few provide details on the process used to convert from a continuous to a discrete scale. Select studies in this survey that provide information about their data preparation process are described below.

In converting from continuous to discrete data, after manually specifying the number of classes (which has been explained in Sect. 4.4), studies automate the selection of the class thresholds using clustering tools, specifically k-means and k-medoids (Chattopadhyay et al., 2020; Mishra and Kushwaha, 2019; Aguasca-

Colomo et al., 2019). Another approach taken is to manually determine suitable thresholds, by performing a series of experiments to compare various threshold values (Shi et al., 2017). To address any resulting class imbalances, researchers may perform random down-sampling to obtain an equal sample distribution across classes as in Aguasca-Colomo et al. (2019), Oswal (2019), and Manandhar et al. (2019).

4.7 Machine Learning Techniques Used

The studies in this survey made use of a wide range of ML techniques which can be subdivided into two main groups: "classical" techniques such as multivariate linear regression (MLR), KNN ANNs, SVMs, and RF; and modern deep learning methods such as CNNs and Long-Short-Term-Memory (LSTM). It was observed that classical ML models tended to work with 1D data from meteorological stations, such as in Mallika and Nirmala (2016), Du et al. (2019), Dash et al. (2018), Amiri et al. (2016), Pham et al. (2019), Yu et al. (2017), and Ramsundram et al. (2016) for short-term data and Du et al. (2019), Cristian (2018), Lakshmaiah et al. (2016), Ramsundram et al. (2016), Abbot and Marohasy (2016), Shenify et al. (2016), Mehr et al. (2019), Damavandi and Shah (2019), Mehdizadeh et al. (2018), Gao et al. (2019), and Aguasca-Colomo et al. (2019) for long-term data.

Some papers use hybrid models that combine two or more approaches. A popular hybrid approach is to combine ML with optimization tools such as genetics and particle swarm optimization to optimize hyper-parameters (Mehdizadeh et al., 2018; Beheshti et al., 2016; Mehr et al., 2019; Du et al., 2018, 2017). Multiple ML techniques are combined in Chhetri et al. (2020), Singh and Kumar (2019), and Peng et al. (2019), and ML is used with ARIMA in Pham et al. (2019).

Deep learning models usually requires huge data sets to avoid overfitting on the data, which explains their popularity among short-term data sets, especially those using 2D data (Shi et al., 2015; Singh et al., 2017; Jing et al., 2019; Sato et al., 2018; Ayzelet al., 2019; Chen et al., 2020; Tran and Song, 2019a; Shi et al., 2017; Castro et al., 2020; Wang et al., 2017; Tran and Song, 2019b; Shi et al., 2017; Pan et al., 2019; Du et al., 2018; Zhan et al., 2019; Chattopadhyay et al., 2020; Patel et al., 2018; Zhuang and Ding, 2016; Boonyuen et al., 2018, 2019). 2D data, in particular, has a huge feature space, which requires authors to implement automated feature reduction models like CNNs (Zhan et al., 2019; Chattopadhyay et al., 2020; Patel et al., 2018; Zhuang and Ding, 2016; Boonyuen et al., 2018).

In order to accommodate the time dimension in the data, many researchers try to adapt time series models such as LSTMs for 1D data in Kumar et al. (2019), Duong et al. (2018), Xu et al. (2020), Aswin et al. (2018), Canchala et al. (2020), Zhang et al. (2020), and Patel et al. (2018). For 2D data, models combining CNNs with LSTMs (designated as ConvLSTMs models) were first used in inShi et al. (2015) in 2015, and subsequently several variations have been implemented (Singh et al.,

2017; Jing et al., 2019; Sato et al., 2018; Ayzelet al., 2019; Chen et al., 2020; Tran and Song, 2019a; Shi et al., 2017; Castro et al., 2020; Wang et al., 2017; Tran and Song, 2019b; Shi et al., 2017).

4.8 Reporting of Results and Accuracy Measures

Several different metrics are used in the literature to measure the performance of the ML models according to the type of the problem. In classification problems, authors tend use metrics such as precision, recall, and accuracy (Gao et al., 2019; Mishra and Kushwaha, 2019; Aguasca-Colomo et al., 2019; Zainudin et al., 2016; Diez-Sierra and del Jesus, 2020; Singh and Kumar, 2019; Oswal, 2019; Chen et al., 2016; Manandhar et al., 2019; Kashiwao et al., 2017; Huang et al., 2017; Singh et al., 2017; Zhan et al., 2019; Chattopadhyay et al., 2020; Patel et al., 2018; Zhuang and Ding, 2016; Boonyuen et al., 2018). If the data is not balanced then f1-score is used rather than the accuracy, since accuracy does not take the imbalance between the classes into account (Zainudin et al., 2016; Diez-Sierra and del Jesus, 2020; Singh and Kumar, 2019; Oswal, 2019; Singh et al., 2017; Patel et al., 2018). For sequence classification prediction, other metrics are used such as the critical success (CSI) (Sato et al., 2018; Ayzelet al., 2019; Chen et al., 2020; Tran and Song, 2019a; Shi et al., 2017; Tran and Song, 2019b; Shi et al., 2017). For continuous outputs, then the mean absolute error, and the root mean squared error are the most commonly used metrics in the literature (Peng et al., 2019; Pham et al., 2019; Valencia-Payan and Corrales, 2018; Zhang et al., 2020, 2018; Yu et al., 2017; Du et al., 2019; Dash et al. , 2018; Cristian, 2018; Lakshmaiah et al., 2016; Ramsundram et al., 2016; Sulaiman and Wahab, 2018; Amiri et al., 2016; Abbot and Marohasy, 2016; Sardeshpande and Thool, 2019; Shenify et al., 2016; Banadkooki et al. , 2019; Mehr et al., 2019; Damavandi and Shah, 2019; Lee et al., 2018; Beheshti et al., 2016; Nourani et al., 2019; Abbot and Marohasy, 2017; Kumar et al., 2019; Haidar and Verma, 2018; Duong et al., 2018; Xu et al., 2020; Aswin et al., 2018; Canchala et al., 2020; Bojang et al., 2020; Chhetri et al., 2020; Mehdizadeh et al., 2018)

A direct comparison of these results across different papers is a nearly impossible task, since each paper uses its own models, pre-processing, metrics, data sets and parameters. However, individual authors frequently compare multiple algorithms, and there are a few ML algorithms that stand out as being most frequently mentioned as better performers. ANNs and deep learning are most frequently mentioned as best performing models, for both long-term prediction (Du et al., 2019; Lakshmaiah et al., 2016; Amiri et al., 2016; Abbot and Marohasy, 2016; Sardeshpande and Thool, 2019; Banadkooki et al. , 2019; Abbot and Marohasy, 2017; Weesakul et al., 2018; Aswin et al., 2018; Kumar et al., 2019; Haidar and Verma, 2018; Duong et al., 2018; Canchala et al., 2020; Chhetri et al., 2020; Mehdizadeh et al., 2018) and especially for short-term prediction (Shi et al., 2015; Singh et al., 2017; Jing et al., 2019; Sato et al., 2018; Ayzelet al., 2019; Chen et al., 2020; Tran and Song, 2019a; Shi et al., 2017; Castro et al., 2020; Wang et al., 2017; Tran and Song, 2019b; Shi et al., 2017;

Pan et al., 2019; Du et al., 2018; Zhan et al., 2019; Chattopadhyay et al., 2020; Patel et al., 2018; Zhuang and Ding, 2016; Boonyuen et al., 2018, 2019; Zhang et al., 2020, 2018; Diez-Sierra and del Jesus, 2020; Kashiwao et al., 2017).

Other algorithms mentioned as best performers are SVMs in 6 studies (Yu et al., 2017; Manandhar et al., 2019; Chen et al., 2016; Du et al., 2017; Shenify et al., 2016; Mehr et al., 2019) ensemble in Valencia-Payan and Corrales (2018), Zainudin et al. (2016), Ramsundram et al. (2016), Nourani et al. (2019), Xu et al. (2020), and Aguasca-Colomo et al. (2019), logistic regression in three studies (Oswal, 2019; Balamurugan and Manojkumar, 2021; Gao et al., 2019) and KNNs in two studies (Huang et al., 2017; Cristian, 2018).

4.9 Discussion

The above sections clearly demonstrate that there is a robust, growing literature on rainfall prediction, which covers an extremely wide variety of time scales, features used, pre-processing techniques, and ML algorithms used. From a high-level perspective, the field can be divided into short versus long time scales (time intervals of a one day or less, versus intervals of a month or more), which tend to have divergent characteristics.

Short-term studies typically rely on huge data sets, and require deep learning applied to large feature sets to find hidden patterns in those data sets. On the other hand, long-term studies rely more on pre-processing methods such as feature selection, data imputation, and data balancing in order to make effective predictions. ANNs and deep learning seem are becoming increasingly prevalent in long-term studies as well as short term: since 2018, 7 of 23 papers on long-term prediction utilized deep learning tools.

There are reasons to regard the trend towards more complicated models with skepticism. Some recent studies have shown that much simpler models such as KNNs can sometimes outperform advanced ML techniques like RNNs, (Lin, 2019; Hussein et al., 2020; Ludewig and Jannach, 2018; Cristian, 2018, Hussein et al., 2021). Similar findings have been reported for other ML applications, such as the top *n* recommendation problem (Dacrema et al., 2019).

These results underscore the importance of providing simple but statistically well-motivated baselines to verify whether ML truly is effective in improving predictive accuracy. However, many papers do not provide simple baselines, but rather compare several variations or architectures of more advanced ML methods such as SVR or MLP (Mohamadi et al., 2020; Bojang et al., 2020; Mehdizadeh et al., 2018; Canchala et al., 2020; Peng et al., 2019; Chen et al., 2016; Du et al., 2017; Manandhar et al., 2019; Du et al., 2018; Zhan et al., 2019; Patel et al., 2018; Boonyuen et al., 2018, 2019). Of the total reviewed papers, almost half (48.2%) of papers did not supply simple baselines. Of those papers that did supply baselines, a variety of methods was used. For short-term image data, the previous image is frequently used as an untrained predictor for the next image (Tran and Song, 2019a;

Shi et al., 2017; Tran and Song, 2019b; Shi et al., 2017). For monthly data, some papers use MLR based on multiple previous lags (Chhetri et al., 2020; Du et al., 2019; Cristian, 2018; Lakshmaiah et al., 2016); while same month averages, though statistically well-motivated, are used much less frequently (Xu et al., 2020).

Besides the issue of baselining, the use of error bars is essential for comparison purposes, as it highlights whether the improvement obtained by the models are significant. Unfortunately most of the literature in ML does not provide error bars around the measured metrics. In the case of our reviewed literature shows that 88% of the papers did not give error bars.

A final issue of concern is data leakage. Data leakage refers to allowing data from the testing set to influence the training set. Data leakage occurs during the pre-processing of the data, and can take various forms as follows:

- Random shuffling, which involves choosing sequences from a common data pool for both training and testing:
- Imputation, which involves filling missing records using statistical methods on the entire data set (including both training and testing)
- De-seasonalization which utilizes the monthly averages from the entire data set.
- Using current lags, e.g., using temperature at a time T to predict rainfall at the same time T. (Depending on the application, this may or may not constitute data leakage)
- Combination: Which uses two of the above-mentioned techniques.

Figure 4.6, shows the reviewed papers in terms of data leakage. The top chart focuses on long-term data, where the bottom focuses on short-term data. We mentioned previously that long-term data often undergoes more pre-processing than short-term data. This reflects on the graph, as leakage-producing methods are more than twice as common for long term as for short term. Random shuffling was performed in Du et al. (2019); Lakshmaiah et al. (2016), Lee et al. (2018), Beheshti et al. (2016), Duong et al. (2018), Gao et al. (2019), Mishra and Kushwaha (2019), and Aguasca-Colomo et al. (2019) for long-term data, and in Valencia-Payan and Corrales (2018), Du et al. (2017), Manandhar et al. (2019), Pan et al. (2019), Du et al. (2018), and Chattopadhyay et al. (2020) for short-term data. Data imputation was performed in Sulaiman and Wahab (2018), Haidar and Verma (2018), Canchala et al. (2020), and Bojang et al. (2020) for long-term data, and in Oswal (2019) for short-term data. Faulty de-seasonalization was carried out in Mehr et al. (2019) for long-term data. Using the current lags was seen only implemented in Ramsundram et al. (2016). Multiple leakage issues (denoted as "combination" in the figure) were observed in Chhetri et al. (2020) and Zainudin et al. (2016).

4.10 Conclusions

In the area of rainfall prediction, 66 relevant papers are reviewed, by examining the data source, output objective, input feature, pre-processing methods, models

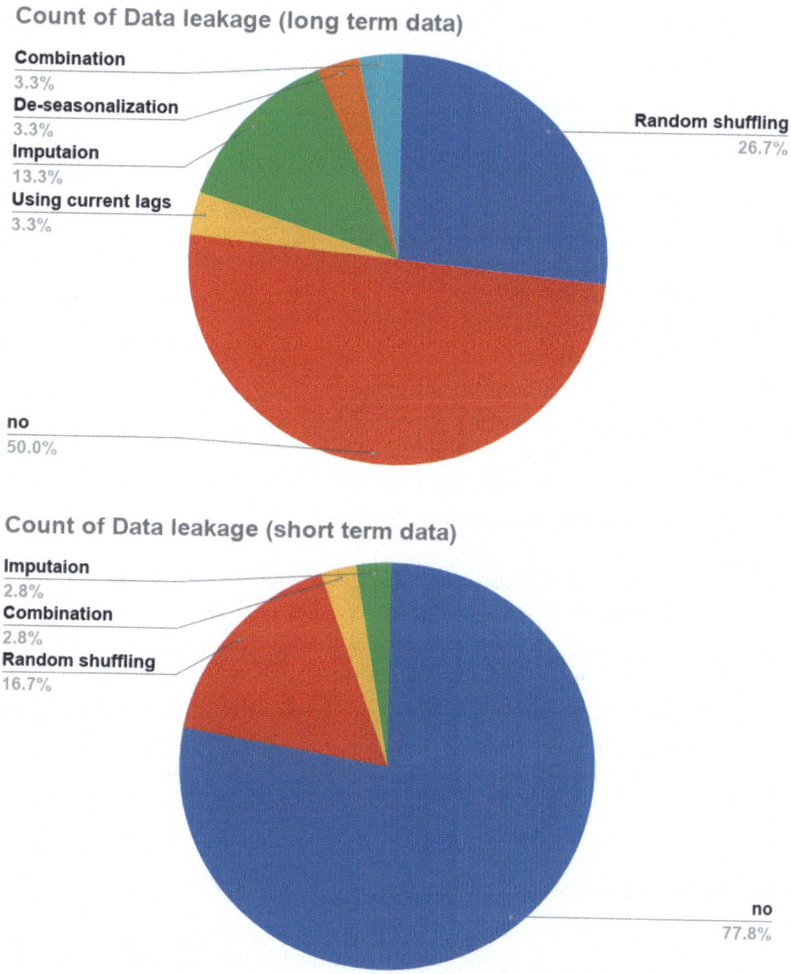

Fig. 4.6 Percentage of papers which introduced data leakage during pre processing, for long-term data (*top*) and short-term data (*bottom*)

used, and finally the results. Different pre-processing like random shuffling used in the literature suggests that in some cases model performance is inaccurately represented. The aim of the survey is to make researches aware of the different pitfalls that can leads to unreal models performance, which does not only apply for rainfall, but for other time series data.

Acknowledgments E.A.H. acknowledges financial support from the South African National Research Foundation (NRF CSUR Grant Number 121291 for the HIPPO project) and from the Telkom-Openserve-Aria Technologies Center of Excellence at the Department of Computer Science of the University of the Western Cape.

Appendix 1: List of Abbreviations

ML	Machine learning
AD	Author defined
ANNs	Artificial neural networks
CNNs	Convolution neural networks
LSTMs	Long short-term memory
ConvLSTMs	Convolutions layers with Long short-term memory
RF	Random forest
RF	SVMs Support vector machines
DT	Decision tress
XGB	Extreme gradient boosting
LogReg	Logistic regression
MLR	Multi linear regression
KNNs	K-nearest neighbour
RMSE	Root mean square error
MAE	mean absolute error
CA	Classification accuracy
pre	precision
f1	f1-score
PACF	Partial autocorrelation function
ACF	Autocorrelation function
PCA	principle component analysis
NOAA	National Oceanic and Atmospheric Administration

Appendix 2: Summary Tables for References

This appendix contains four tables which summarize the findings for the reviewed papers for long-term data Tables 4.1 and 4.2, and short-term data Tables 4.3 and 4.4. Tables 4.1 and 4.3 contain information regarding the source, period, region, input, output; while Tables 4.2 and4.4 include information about the pre-processing tools, data leakage, and the ML used.

Table 4.1 Data sources, spatio-temporal coverage, inputs and outputs, and references for long-term predictive studies

No.	Source	Period	Region	Input	Output	Ref
1	China meteorological administration (CMA)	1916–2015	China	6 climatic indices	Seasonal regression	Du et al. (2019)
2	Indian institute of tropical meteorology (IITM)	1817–2016	India	8 past lags	Seasonal regression	Dash et al. (2018)
3	Romanian rainfall	1991–2015	Romania	12 past lags	Monthly regression	Cristian (2018)
4	Rainfall from the India water portal	1901–2002	India	11 climatic parameters	Monthly regression	Lakshmaiah et al. (2016)
5	Tuticorin meteorological station	1980–2002	India	Four climatic parameters	Monthly regression	Ramsundram et al. (2016)
6	Malaysian department of irrigation and drainage	1965–2015	Malaysia	10 past lags	Monthly regression	Sulaiman and Wahab (2018)
7	National cartographic center of Iran (NCC)	1996–2010	Iran	Four climatic parameters	Monthly regression	Amiri et al. (2016)
8	Royal Netherlands meteorological institute climate explorer	2004–2014	Australia	Seven climatic indices	Monthly regression	Abbot and Marohasy (2016)
9	Indian water portal	1901–2000	India	Four climatic parameters	Monthly regression	Sardeshpande and Thool (2019)
10	Serbian meteorological stations	1946–2012	Serbia	Past rainfall lags	Monthly regression	Shenify et al. (2016)
11	Iran meteorological department	2000–2010	Iran	Two climatic parameters	Monthly regression	Banadkooki et al. (2019)
12	Iran meteorological department	1990–2014	Iran	Four past lags	Monthly regression	Mehr et al. (2019)
13	CHIRPS, and NCEP-NCAR Reanalysis	1918–2001	Indus basin	5 climatic features	Monthly regression	Damavandi and Shah (2019)
14	World agrometeorological information service (WAMIS) and NOAA	1966–2017	South Korea	11 climatic indices	Monthly regression	Lee et al. (2018)
15	Malaysian department of irrigation and drainage	1950–2010	Malaysia	6 past lags and time stamp	Monthly regression	Beheshti et al. (2016)
16	Turkish stations	2007–2016	Turkey	3 rainfall lags	Monthly regression	Nourani et al. (2019)

(continued)

Table 4.1 (continued)

No.	Source	Period	Region	Input	Output	Ref
17	Australian stations	1885–2014	Australia	10 climatic indices and parameters	Monthly regression	Abbot and Marohasy (2017)
18	Indian meteorological department	1871–2016	India	12 past lags	Monthly regression	Kumar et al. (2019)
19	Bureau of meteorology (BOM), Royal Netherlands meteorological institute climate, more	1908–2012	Australia	43 climatic indices and parameters	Monthly regression	Haidar and Verma (2018)
20	Vietnam's hydrological gauging	1971–2010	Vietnam	12 features	Monthly regression	Duong et al. (2018)
21	Global precipitation climatology center (GPCC)	1901–2013	China	6–9 climatic indices and parameters	Monthly regression	Xu et al. (2020)
22	Precipitation from NCEP	1979–2018	GLOBAL	164 past lags	Monthly regression	Aswin et al. (2018)
23	National center of hydrology and meteorology department (NCHM)	1997–2015	Bhutan	6 climates parameters	Monthly regression	Canchala et al. (2020)
24	Taiwan water resource bureau	1958–2018	Taiwan	3 past lag	Monthly regression	Bojang et al. (2020)
25	Instituto de Hidrología, Meterología y Estudios Ambientales (IDEAM) of Colombia	1983–2016	Colombia	6 past lags	Monthly regression	Chhetri et al. (2020)
26	Islamic Republic of Iran meteorological organization (IRIMO)	1981- 2012	Iran	5 past lags	Monthly regression	Mehdizadeh et al. (2018)
27	Pluak Daeng station in Thailand	1991–2016	Thailand	346 climatic indices and parameters	Monthly regression	Weesakul et al. (2018)
28	National climate center of China meteorological administration (NCC-CMA)	1952–2012	China	84 climatic indices	Yearly Classification	Gao et al. (2019)
29	The department of agricultural meteorology Indira	2011–2013	India	Five climatic parameters	Monthly classification	Mishra and Kushwaha (2019)
30	Meteorological stations of the island of Tenerife and NOAA databases	1976–2016	Tenerife Island	12 climatic indices and parameters	Monthly classification	Aguasca-Colomo et al. (2019)

Table 4.2 Pre-processing, data leakage characteristics, machine learning algorithms used, and reference numbers for long-term predictive studies

No.	Pre-processing	Data leakage	ML used	Ref
1	Normalization, random shuffling, feature correlation	Random shuffling	PCA-ANN, PCA-MLR	Du et al. (2019)
2	Normalization	No	KNNs, ANNs, ELM	Dash et al. (2018)
3	Windowing	No	KNNs, ARIMA, ANNs	Cristian (2018)
4	Windowing, random shuffling	Random shuffling	ANN, ARMA, LR	Lakshmaiah et al. (2016)
5	Data imputation, noise removal, correlation analysis	Using current lags	DT, ANNs	Ramsundram et al. (2016)
6	Normalization, and data imputation	Imputation	ANNs, ARIMA	Sulaiman and Wahab (2018)
7	Normalization, decomposition	no	WTANN, ANNs	Amiri et al. (2016)
8	Features correlation	No	ANNs, POAMA	Abbot and Marohasy (2016)
9	Normalization	No	Different ANNs	Sardeshpande and Thool (2019)
10	N/A	No	ANN, WT-SVM, GP	Shenify et al. (2016)
11	Normalization, optimization	No	AD-MLP, AD-SVM, DT	Banadkooki et al. (2019)
12	Correlation analysis (PACF), square root transformation, standardization, de-seasonalization	De-seasonalization	SVR, AD-SVR, more	Mehr et al. (2019)
13	Feature correlation, random shuffling	No	MLP, SVR, MLR, RF, KNNs	Damavandi and Shah (2019)
14	Feature correlation, random shuffling	Random shuffling	ANNs	Lee et al. (2018)
15	Decomposition	Random shuffling	AD-MLP	Beheshti et al. (2016)
16	Normalization	No	Ensemble method, SVM, ANNS, more	Nourani et al. (2019)
17	Feature selection	No	ANNs, POAMA	Abbot and Marohasy (2017)
18	Feature correlation, windowing	N/A	LSTM, RNN	Kumar et al. (2019)
19	Data imputation, normalization	Imputation	1D-CNN, MLP, baseline (ACCESS-S1)	Haidar and Verma (2018)

(continued)

Table 4.2 (continued)

No.	Pre-processing	Data leakage	ML used	Ref
20	Random shuffling	Random shuffling	MLP, LSTM, SNN	Duong et al. (2018)
21	Normalization, wavelet	No	MLR, MLP, LSTM, SVMs, ConvLSTMs, ensemble methods	Xu et al. (2020)
22	Grey-scale, windowing	No	LSTM, ConvNet	Aswin et al. (2018)
23	Normalization, data imputation	Imputation	MLR , AD-LSTM, LSTM, MLP	Canchala et al. (2020)
24	Decomposition	Imputation	UD-RF, RF, UD-SVR, SVR	Bojang et al. (2020)
25	Imputation, de-seasonlization	Imputation, de-seasonalization	3 AD-ANNs models	Chhetri et al. (2020)
26	Normalization	No	ANNs, AD-ANNs, AD-gene expression programming	Mehdizadeh et al. (2018)
27	N/A	No	DNNs	Weesakul et al. (2018)
28	Feature correlation, feature reduction	Random shuffling	MLogR	Gao et al. (2019)
29	Clustering	Random shuffling	GPR, DT, NB	Mishra and Kushwaha (2019)
30	Random down-sampling, feature correlation	Random shuffling	XGB, RF, more	Aguasca-Colomo et al. (2019)

Table 4.3 Data sources, spatio-temporal coverage, inputs and outputs, and references for short-term predictive studies

No.	Source	Period	Region	Input	Output	Ref
1	Indian statistical institute	1939–1995	Multiple regions	10 climatic parameters	Daily regression	Peng et al. (2019)
2	Vietnamese stations	1978–2016	Vietnam	Previous lags	Daily regression	Pham et al. (2019)
3	Meteoblue data , MODIS, and more	2012–2014	Colombia	12 climatic indices and parameters	Hourly regression	Valencia-Payan and Corrales (2018)
4	Central meteorological observatory of Shanghai	2015–2017	China	24 climatic parameters	Hourly regression	Zhang et al. (2020)
5	China meteorological administration	2015–2017	China	13 climatic parameters	Hourly regression	Zhang et al. (2018)
6	Taiwan and the national severe storms laboratory and NOAA	2012–2015	Taiwan	3–4 parameters	Hourly regression	Yu et al. (2017)
7	Meteorological drainage and the irrigation departments in Malaysia	2010–2014.	Malaysia	4 parameters	Daily classification	Zainudin et al. (2016)
8	The water planning and managing agency for Tenerife Island, and NOAA	1979–2015	Spain	1800 parameters	Daily classification	Diez-Sierra and del Jesus (2020)
9	U.S. government's open data	2010–2017	US	25 parameters	Daily classification	Singh and Kumar (2019)
10	Kaggle and the Australian government	2008–2017	Australia	23 parameters	Daily classification	Oswal (2019)
11	Indian meteorological department	2008–2017	India	8 parameters	Daily classification	Balamurugan and Manojkumar (2021)
12	Satellite imagery data are from FY-2G, and meteorological station located in Shenzhen	2015	China	8 parameters	Hourly classification	Chen et al. (2016)

(continued)

Table 4.3 (continued)

No.	Source	Period	Region	Input	Output	Ref
13	Data from the Nanjing station	N/A	China	6 parameters	Hourly classification	Du et al. (2017)
14	Singapore related weather stations	2012–2015	Singapore	15 climatic parameters	Min classification	Manandhar et al. (2019)
15	Japan meteorological agency	2000–2012	Japan	8 features	Min classification	Kashiwao et al. (2017)
16	NCEP-NCAR and Beijing meteorological station	1990–2012	China	6 climatic indices and parameters	Daily classification	Huang et al. (2017)
17	Radar images collected in Hong Kong	2011–2013	Hong Kong	5 frames	Min classification	Shi et al. (2015)
18	Radar images from USA from 2008–2015	2008–2015	US	10 frames	Min classification	Singh et al. (2017)
19	Radar images from national meteorological information center	2016–2017	China	10 frames	Min classification	Jing et al. (2019)
20	Radar images are retrieved using Yahoo! static map API	2013–2017	Japan	10 frames	Min classification	Sato et al. (2018)
21	Radar images from the German weather service (DWD)	2006–2017	Germany	2 frames	Min both	Ayzelet al. (2019)
22	Weather surveillance radar-1988 doppler radar (WSR-88D)	2015–2018	China	20 frames	Min classification	Chen et al. (2020)
23	CIKM AnalytiCup 2017 competition	N/A	China	5 frames	Min regression	Tran and Song (2019a)
24	CINRAD-SA type Doppler weather radar	2016	China	4 frames	Min classification	Shi et al. (2017)
25	CHIRPS	1918–2019	China	5 frames	Daily regression	Castro et al. (2020)

26	Radar images collected in Hong Kong	2011–2013	China	10 frames	Min regression	Wang et al. (2017)
27	CIKM AnalytiCup 2017 competition	N/A	China	7 frames	Min both	Tran and Song (2019b)
28	Dataset from HKO-7	2009–2015	Hong Kong	5 frames	Min regression	Shi et al. (2017)
29	NCEP, and NOAA	1979–2017	US	A tensor of $8 \times 4 \times 25 \times 25$	Daily regression	Pan et al. (2019)
30	China meteorological data network	N/A	China	7 climatic parameters	Hourly regression	Du et al. (2018)
31	NOAA	1800–2017	US	30 climatic parameters	Hourly both	Zhan et al. (2019)
32	Large ensemble (LENS) community project	1920–2005	US	$3 \times 28 \times 28 \times 3$	Hourly classification	Chattopadhyay et al. (2020)
33	Kaggle	2012–2017	US and India	120 climatic lags	Hourly classification	Patel et al. (2018)
34	Iowa state	1948–2010	USA	9 climatic parameters	Daily classification	Zhuang and Ding (2016)
35	Meteorological department of Thailand and the petroleum authority of Thailand	2017–2017	Thailand	One image	Daily classification	Boonyuen et al. (2018)
36	Meteorological department of Thailand and the petroleum authority of Thailand	2017–2018	Thailand	One and batch of images	Daily classification	Boonyuen et al. (2019)

Table 4.4 Pre-processing, data leakage characteristics, machine learning algorithms used, and reference numbers for short-term predictive studies

No.	Pre-processing	Data leakage	ML used	Ref
1	Normalization, cross validation, feature reduction (PCA)	No	AD-ELM	Peng et al. (2019)
2	Normalization, feature correlation	No	ARIMA-MLP, ARIMA-SVM, ARIMA-HW, ARIMA-NF, more	Pham et al. (2019)
3	Data imputation, data shuffling	Random shuffling	RF, cubist	Valencia-Payan and Corrales (2018)
4	Feature selection, correlation analysis, interpolation, clustering	No	LSTM, MLR, SVMs, ECMFWF	Zhang et al. (2020)
5	Normalization, feature reduction (PCA)	No	DRCF, ARIMA, more	Zhang et al. (2018)
6	N/A	No	RF, SVM	Yu et al. (2017)
7	Normalization, data imputation, shuffling	Data imputation, random shuffling	SVM, RF, DT, NB, ANN	Zainudin et al. (2016)
8	Feature reduction (PCA)	No	ANNs, RF, KNNs, LogR	Diez-Sierra and del Jesus (2020)
9	Feature selection (RF), k-fold cross validation	No	RF,AD[ANNs, Adaboost, SVM, KNN]	Singh and Kumar (2019)
10	Feature selection, feature correlation, data imputation, over, and down-sampling	Imputation	LogR, DT, KNNs, more	Oswal (2019)
11	N/A	No	LogReg, DT, RF, more	Balamurugan and Manojkumar (2021)
12	Radiometric, and geometric correction, and windowing	No	SVM	Chen et al. (2016)
13	Normalization, random shuffling	Random shuffling	AD-SVMs	Du et al. (2017)
14	Down-sampling, feature correlation	Random shuffling	SVM	Manandhar et al. (2019)

15	Outliers removal, normalization	No	MLP, RBFN	Kashiwao et al. (2017)
16	Normalization	No	Knns	Huang et al. (2017)
17	Feature reduction, noise removal, windowing	No	ConvLSTM, FC-LSTM, more	Shi et al. (2015)
18	Resizing, windowing	No	Eulerian persistence, AD-Conv-RNN, ConvLSTM	Singh et al. (2017)
19	Feature reduction, windowing	No	MLC-LSTM, ConvLSTM, more	Jing et al. (2019)
20	Feature reduction, windowing	No	SDPredNet, TrajGRU, more	Sato et al. (2018)
21	Logarithmic transformation	No	Optical flow, Dozhdya.Net	Ayzelet al. (2019)
22	Noise removal, remove corrupted images, windowing, Normalization	No	COTREC, ConvLSTM, AD-ConvLSTM, more	Chen et al. (2020)
23	Normalization, windowing	No	Last frame, TrajGRU, ConvLSTM, AD-TrajGRU, more	Tran and Song (2019a)
24	Windowing, grey-scale transformation	No	Last input, COTREC, AD-CNN	Shi et al. (2017)
25	Windowing, grey-scale, resizing	No	ConvLSTM, AD-ConvLSTMs	Castro et al. (2020)
26	Windowing, grey-scale, resizing	No	ConvLSTM, PredRNN, VPNbaseline	Wang et al. (2017)
27	Windowing, grey-scale, resizing, data augmentation	No	ConvLSTM, ConvGRU, TrajGRU, PredRNN, PredRNN++, last frame	Tran and Song (2019b)
28	Windowing, grey-scale, noise removal, normalization	No	2D CNN, 3D CNN, ConvGRU, TrajGRU, last frame, more	Shi et al. (2017)
29	Normalization, random shuffling	Random shuffling	LR, CNNs, base model (NARR)	Pan et al. (2019)
30	Random shuffling and normalization	Random shuffling	DBN, GA-SVM, more	Du et al. (2018)
31	N/A	No	CNN, LPBoost, more	Zhan et al. (2019)
32	Clustering, down-sampling, random shuffling	Random shuffling	CNN, LogReg	Chattopadhyay et al. (2020)
33	Normalization	No	CNN, LSTM	Patel et al. (2018)
34	Cropping	No	CNN	Zhuang and Ding (2016)
35	Cropping	N/A	CNN	Boonyuen et al. (2018)
36	Cropping	N/A	CNN	Boonyuen et al. (2019)

References

Abbot, J., & Marohasy, J. (2016). Forecasting monthly rainfall in the western Australian wheat-belt up to 18-months in advance using artificial neural networks. In *Australasian Joint Conference on Artificial Intelligence* (pp. 71–87). Berlin: Springer.

Abbot, J., & Marohasy, J. (2017). Application of artificial neural networks to forecasting monthly rainfall one year in advance for locations within the Murray Darling basin, Australia. *International Journal of Sustainable Development and Planning, 12*(8), 1282–1298.

Aguasca-Colomo, R., Castellanos-Nieves, D., & Méndez, M. (2019). Comparative analysis of rainfall prediction models using machine learning in islands with complex orography: Tenerife island. *Applied Sciences, 9*(22), 4931.

Amiri, M. A., Amerian, Y., & Mesgari, M. S. (2016). Spatial and temporal monthly precipitation forecasting using wavelet transform and neural networks, Qara-Qum catchment, Iran. *Arabian Journal of Geosciences, 9*(5), 421.

Aswin, S., Geetha, P., & Vinayakumar, R. (2018). Deep learning models for the prediction of rainfall. In *2018 International Conference on Communication and Signal Processing (ICCSP)* (pp. 0657–0661). Piscataway: IEEE.

Ayzel, G., Heistermann, M., Sorokin, A., Nikitin, O., & Lukyanova, O. (2019). All convolutional neural networks for radar-based precipitation nowcasting. *Procedia Computer Science, 150*, 186–192.

Balamurugan, M. S., & Manojkumar, R. (2021). Study of short term rain forecasting using machine learning based approach. *Wireless Networks, 27*, 5429–5434.

Banadkooki, F. B., Ehteram, M., Ahmed, A. N., Fai, C. M., Afan, H. A., Ridwam, W. M., Sefelnasr, A., & El-Shafie, A. (2019). Precipitation forecasting using multilayer neural network and support vector machine optimization based on flow regime algorithm taking into account uncertainties of soft computing models. *Sustainability, 11*(23), 6681.

Barnett, A. G., Baker, P., & Dobson, A. (2012). Analysing seasonal data. *R Journal, 4*(1), 5–10.

Beheshti, Z., Firouzi, M., Shamsuddin, S. M., Zibarzani, M., & Yusop, Z. (2016). A new rainfall forecasting model using the CAPSO algorithm and an artificial neural network. *Neural Computing and Applications, 27*(8), 2551–2565.

Bojang, P. O., Yang, T.-C., Pham, Q. B., & Yu, P.-S. (2020). Linking singular spectrum analysis and machine learning for monthly rainfall forecasting. *Applied Sciences, 10*(9), 3224.

Boonyuen, K., Kaewprapha, P., & Srivihok, P. (2018). Daily rainfall forecast model from satellite image using convolution neural network. In *2018 IEEE International Conference on Information Technology* (pp. 1–7).

Boonyuen, K., Kaewprapha, P., Weesakul, U., & Srivihok, P. (2019). Convolutional neural network inception-v3: A machine learning approach for leveling short-range rainfall forecast model from satellite image. In *International Conference on Swarm Intelligence* (pp. 105–115). Berlin: Springer.

Canchala, T., Alfonso-Morales, W., Carvajal-Escobar, Y., Cerón, W. L., & Caicedo-Bravo, E. (2020). Monthly rainfall anomalies forecasting for southwestern Colombia using artificial neural networks approaches. *Water, 12*(9), 2628.

Castro, R., Souto, Y. M., Ogasawara, E., Porto, F., & Bezerra, E. (2020). STConvS2S: Spatiotemporal convolutional sequence to sequence network for weather forecasting. *Neurocomputing, 426*, 285–298.

Chattopadhyay, A., Hassanzadeh, P., & Pasha, S. (2020). Predicting clustered weather patterns: A test case for applications of convolutional neural networks to spatio-temporal climate data. Sci. Rep. **10**(1), 1–13.

Chen, L., Cao, Y., Ma, L., & Zhang, J. (2020). A deep learning based methodology for precipitation nowcasting with radar. *Earth and Space Science, 7*, e2019EA000812.

Chen, K., Liu, J., Guo, S., Chen, J., Liu, P., Qian, J., Chen, H., & Sun, B. (2016). Short-term precipitation occurrence prediction for strong convective weather using fy2-g satellite data: A case study of Shenzhen, South China. *The International Archives of Photogrammetry, Remote Sensing and Spatial Information Sciences, 41*, 215.

Chhetri, M., Kumar, S., Pratim Roy, P., & Kim, B.-G. (2020). Deep BLSTM-GRU model for monthly rainfall prediction: A case study of Simtokha, Bhutan. *Remote Sensing, 12*(19), 3174.

Cristian, M. (2018). Average monthly rainfall forecast in Romania by using k-nearest neighbors regression. *Analele Universității Constantin Brâncuşi din Târgu Jiu: Seria Economie, 1*(4), 5–12.

Dacrema, M. F., Cremonesi, P., & Jannach, D. (2019). Are we really making much progress? A worrying analysis of recent neural recommendation approaches. In *Proceedings of the 13th ACM Conference on Recommender Systems* (pp. 101–109).

Damavandi, H. G., & Shah, R. (2019). A learning framework for an accurate prediction of rainfall rates. arXiv:1901.05885.

Dash, Y., Mishra, S. K., & Panigrahi, B. K. (2018). Rainfall prediction for the Kerala state of India using artificial intelligence approaches. *Computers & Electrical Engineering, 70*, 66–73.

Delleur, J. W., & Kavvas, M. L. (1978). Stochastic models for monthly rainfall forecasting and synthetic generation. *Journal of Applied Meteorology, 17*(10), 1528–1536.

Diez-Sierra, J., & del Jesus, M. (2020). Long-term rainfall prediction using atmospheric synoptic patterns in semi-arid climates with statistical and machine learning methods. *Journal of Hydrology, 586*, 124789.

Du, Y., Berndtsson, R., An, D., Zhang, L., Yuan, F., Uvo, C. B., & Hao, Z. (2019). Multi-space seasonal precipitation prediction model applied to the source region of the Yangtze river, China. *Water, 11*(12), 2440.

Du, J., Liu, Y., & Liu, Z. (2018). Study of precipitation forecast based on deep belief networks. *Algorithms, 11*(9), 132.

Du, J., Liu, Y., Yu, Y., & Yan, W. (2017). A prediction of precipitation data based on support vector machine and particle swarm optimization (PSO-SVM) algorithms. *Algorithms, 10*(2), 57.

Duong, T. A., Bui, M. D., & Rutschmann, P. (2018). A comparative study of three different models to predict monthly rainfall in Ca Mau, Vietnam. In *Wasserbau-Symposium Graz 2018. Wasserwirtschaft–Innovation aus Tradition. Tagungsband. Beiträge zum 19. Gemeinschafts-Symposium der Wasserbau-Institute TU München, TU Graz und ETH Zürich* (p. Paper–G5).

Gao, L., Wei, F., Yan, Z., Ma, J., & Xia, J. (2019). A study of objective prediction for summer precipitation patterns over eastern China based on a multinomial logistic regression model. *Atmosphere, 10*(4), 213.

Haidar, A., & Verma, B. (2018). Monthly rainfall forecasting using one-dimensional deep convolutional neural network. *IEEE Access, 6*, 69053–69063.

Htike, K. K., & Khalifa, O. O. (2010). Rainfall forecasting models using focused time delay neural networks. In *International Conference on Computer and Communication Engineering (ICCCE'10)* (pp. 1–6). Piscataway: IEEE.

Huang, M., Lin, R., Huang, S., & Xing, T. (2017). A novel approach for precipitation forecast via improved k-nearest neighbor algorithm. *Advanced Engineering Informatics, 33*, 89–95.

Hung, N. Q., Babel, M. S., Weesakul, S., & Tripathi, N. K. (2009). An artificial neural network model for rainfall forecasting in Bangkok, Thailand. *Hydrology and Earth System Sciences, 13*(8), 1413–1425.

Hussein, E., Ghaziasgar, M., & Thron, C. (2020). Regional rainfall prediction using support vector machine classification of large-scale precipitation maps. In *2020 IEEE 23rd International Conference on Information Fusion (FUSION)* (pp. 1–8). Piscataway: IEEE.

Hussein, E. A., Ghaziasgar, M., Thron, C., Vaccari, M., & Bagula, A. (2021). Basic statistical estimation outperforms machine learning in monthly prediction of seasonal climatic parameters. *Atmosphere, 12*(5), 539.

Jing, J., Li, Q., & Peng, X. (2019). MLC-LSTM: Exploiting the spatiotemporal correlation between multi-level weather radar echoes for echo sequence extrapolation. *Sensors, 19*(18), 3988.

Karimi, H. A. (2014). *Big data: Techniques and technologies in geoinformatics*. Boca Raton: CRC Press.

Kashiwao, T., Nakayama, K., Ando, S., Ikeda, K., Lee, M., & Bahadori, A. (2017). A neural network-based local rainfall prediction system using meteorological data on the internet: A case study using data from the Japan meteorological agency. *Applied Soft Computing, 56*, 317–330.

Kumar, D., Singh, A., Samui, P., & Jha, R. K. (2019). Forecasting monthly precipitation using sequential modelling. *Hydrological Sciences Journal, 64*(6), 690–700.

Lakshmaiah, K., Murali Krishna, S., & Eswara Reddy, B. (2016). Application of referential ensemble learning techniques to predict the density of rainfall. In *2016 International Conference on Electrical, Electronics, Communication, Computer and Optimization Techniques (ICEECCOT)* (pp. 233–237). Piscataway: IEEE.

Lee, J., Kim, C.-G., Lee, J. E., Kim, N. W., & Kim, H. (2018). Application of artificial neural networks to rainfall forecasting in the Geum river basin, Korea. *Water, 10*(10), 1448.

Lin, J. (2019). The neural hype and comparisons against weak baselines. In *ACM SIGIR forum* (vol. 52, pp. 40–51). New York: ACM.

Lu, J., Hu, W., & Zhang, X. (2018). Precipitation data assimilation system based on a neural network and case-based reasoning system. *Information, 9*(5), 106.

Ludewig, M., & Jannach, D. (2018). Evaluation of session-based recommendation algorithms. *User Modeling and User-Adapted Interaction, 28*(4–5), 331–390.

Mallika, M., & Nirmala, M. (2016). Chennai annual rainfall prediction using k-nearest neighbour technique. *International Journal of Pure and Applied Mathematics, 109*(8), 115–120.

Manandhar, S., Dev, S., Lee, Y. H., Meng, Y. S., & Winkler, S. (2019). A data-driven approach for accurate rainfall prediction. *IEEE Transactions on Geoscience and Remote Sensing, 57*(11), 9323–9331.

Mehdizadeh, S., Behmanesh, J., & Khalili, K. (2018). New approaches for estimation of monthly rainfall based on GEP-ARCH and ANN-ARCH hybrid models. *Water Resources Management, 32*(2), 527–545.

Mehr, A. D., Nourani, V., Khosrowshahi, V. K., & Ghorbani, M. A. (2019). A hybrid support vector regression–firefly model for monthly rainfall forecasting. *International Journal of Environmental Science and Technology, 16*(1), 335–346.

Mishra, N., & Kushwaha, A. (2019). Rainfall prediction using gaussian process regression classifier. *International Journal of Advanced Research in Computer Engineering & Technology (IJARCET), 8*(8), 392–397.

Mohamadi, S., Ehteram, M., & El-Shafie, A. (2020). Accuracy enhancement for monthly evaporation predicting model utilizing evolutionary machine learning methods. *International Journal of Environmental Science and Technology, 17*, 1–24.

Mosavi, A., Ozturk, P., & Chau, K.-W. (2018). Flood prediction using machine learning models: Literature review. *Water, 10*(11), 1536.

Nasseri, M., Asghari, K., & Abedini, M. J. (2008). Optimized scenario for rainfall forecasting using genetic algorithm coupled with artificial neural network. *Expert Systems with Applications, 35*(3), 1415–1421.

Nielsen, A. (2020). *Practical time series analysis: Prediction with statistics and machine learning*. Sebastopol: O'Reilly.

Nourani, V., Uzelaltinbulat, S., Sadikoglu, F., & Behfar, N. (2019). Artificial intelligence based ensemble modeling for multi-station prediction of precipitation. *Atmosphere, 10*(2):80.

Oswal, N. (2019). Predicting rainfall using machine learning techniques. arXiv:1910.13827.

Pan, B., Hsu, K., AghaKouchak, A., & Sorooshian, S. (2019). Improving precipitation estimation using convolutional neural network. *Water Resources Research, 55*(3), 2301–2321.

Pantanowitz, A., & Marwala, T. (2009). Missing data imputation through the use of the random forest algorithm. In *Advances in Computational Intelligence* (pp. 53–62). Berlin: Springer.

Parmar, A., Mistree, K., & Sompura, M. (2017). Machine learning techniques for rainfall prediction: A review. In *International Conference on Innovations in Information Embedded and Communication Systems*.

Patel, M., Patel, A., Ghosh, R. (2018). Precipitation nowcasting: Leveraging bidirectional LSTM and 1d CNN. arXiv:1810.10485.

Peng, Y., Zhao, H., Zhang, H., Li, W., Qin, X., Liao, J., Liu, Z., Li, J. (2019). An extreme learning machine and gene expression programming-based hybrid model for daily precipitation prediction. *International Journal of Computational Intelligence Systems, 12*(2), 1512–1525.

Pham, Q. B., Abba, S. I., Usman, A. G., Linh, N. T. T., Gupta, V., Malik, A., Costache, R., Vo, N. D., & Tri, D. Q. (2019). Potential of hybrid data-intelligence algorithms for multi-station modelling of rainfall. *Water Resources Management, 33*(15), 5067–5087.

Ramsundram, N., Sathya, S., & Karthikeyan, S. (2016). Comparison of decision tree based rainfall prediction model with data driven model considering climatic variables. *Irrigation and Drainage Systems Engineering, 5*(3).

Sardeshpande, K. D., & Thool, V. R. (2019). Rainfall prediction: A comparative study of neural network architectures. In *Emerging Technologies in Data Mining and Information Security* (pp. 19–28). Berlin: Springer.

Sato, R., Kashima, H., & Yamamoto, T. (2018). Short-term precipitation prediction with skip-connected PredNET. In *International Conference on Artificial Neural Networks* (pp. 373–382). Berlin: Springer.

Shah, A. D., Bartlett, J. W., Carpenter, J., Nicholas, O., & Hemingway, H. (2014). Comparison of random forest and parametric imputation models for imputing missing data using mice: A caliber study. *American Journal of Epidemiology, 179*(6), 764–774.

Shenify, M., Danesh, A. S., Gocić, M., Taher, R. S., Wahab, Ainuddin, W. A., Gani, A., Shamshirband, S., & Petković, D. (2016). Precipitation estimation using support vector machine with discrete wavelet transform. *Water Resources Management, 30*(2), 641–652.

Shi, E., Li, Q., Gu, D., & Zhao, Z. (2017). Convolutional neural networks applied on weather radar echo extrapolation. In *DEStech Transactions on Computer Science and Engineering* (case), 695–704. DEStech Publications.

Shi, X., Chen, Z., Wang, H., Yeung, D., Wong, W., & Woo, W. C. (2015). Convolutional LSTM network: A machine learning approach for precipitation nowcasting. ArXiv, abs/1506.04214.

Shi, X., Gao, Z., Lausen, L., Wang, H., Yeung, D.-Y., Wong, W., Woo, W.-C. (2017). Deep learning for precipitation nowcasting: A benchmark and a new model . In *Advances in Neural Information Processing Systems* (pp. 5617–5627).

Shi, X., & Yeung, D.-Y. (2018). Machine learning for spatiotemporal sequence forecasting: A survey. arXiv:1808.06865.

Singh, G., & Kumar, D. (2019). Hybrid prediction models for rainfall forecasting. In *2019 9th International Conference on Cloud Computing, Data Science & Engineering (Confluence)* (pp. 392–396). Piscataway: IEEE.

Singh, S., Sarkar, S., & Mitra, P. (2017). Leveraging convolutions in recurrent neural networks for doppler weather radar echo prediction. In *International Symposium on Neural Networks* (pp. 310–317). Berlin: Springer.

Sulaiman, J., & Wahab, S. H. (2018). Heavy rainfall forecasting model using artificial neural network for flood prone area. In *IT Convergence and Security 2017* (pp. 68–76). Berlin: Springer.

Tang, F., & Ishwaran, H. (2017). Random forest missing data algorithms. *Statistical Analysis and Data Mining: The ASA Data Science Journal, 10*(6), 363–377.

Tran, Q.-K., & Song, S.-K. (2019a). Computer vision in precipitation nowcasting: Applying image quality assessment metrics for training deep neural networks. *Atmosphere, 10*(5), 244.

Tran, Q.-K., & Song, S.-K. (2019b). Multi-channel weather radar echo extrapolation with convolutional recurrent neural networks. *Remote Sensing, 11*(19), 2303.

Valencia-Payan, C., & Corrales, J. C. (2018). A rainfall prediction tool for sustainable agriculture using random forest. In *Mexican International Conference on Artificial Intelligence* (pp. 315–326). Berlin: Springer.

Wang, Y., Long, M., Wang, J., Gao, Z., & Philip, S. Y. (2017). PredRNN: Recurrent neural networks for predictive learning using spatiotemporal LSTMs. In *Advances in Neural Information Processing Systems* (pp. 879–888).

Weesakul, U., Kaewprapha, P., Boonyuen, K., & Mark, O. (2018). Deep learning neural network: A machine learning approach for monthly rainfall forecast, case study in eastern region of Thailand. *Engineering and Applied Science Research, 45*(3), 203–211.
Xu, L., Chen, N., Zhang, X., & Chen, Z. (2020). A data-driven multi-model ensemble for deterministic and probabilistic precipitation forecasting at seasonal scale. *Climate Dynamics, 54*, 3355–3374.
Yu, P.-S., Yang, T.-C., Chen, S.-Y., Kuo, C.-M., & Tseng, H.-W. (2017). Comparison of random forests and support vector machine for real-time radar-derived rainfall forecasting. *Journal of Hydrology, 552*, 92–104.
Zainudin, S., Jasim, D. S., & Bakar, A. A. (2016). Comparative analysis of data mining techniques for Malaysian rainfall prediction. *International Journal on Advanced Science, Engineering and Information Technology, 6*(6), 1148–1153.
Zhan, C., Wu, F., Wu, Z., & Chi, K. T. (2019). Daily rainfall data construction and application to weather prediction. In *2019 IEEE International Symposium on Circuits and Systems (ISCAS)* (pp. 1–5). Piscataway: IEEE.
Zhang, C.-J., Zeng, J., Wang, H.-Y., Ma, L.-M., & Chu, H. (2020). Correction model for rainfall forecasts using the LSTM with multiple meteorological factors. *Meteorological Applications, 27*(1), e1852.
Zhang, P., Jia, Y., Gao, J., Song, W., & Leung, H. K. N. (2018). Short-term rainfall forecasting using multi-layer perceptron. *IEEE Transactions on Big Data, 6*, 93–106.
Zhuang, W. Y., & Ding, W. (2016). Long-lead prediction of extreme precipitation cluster via a spatiotemporal convolutional neural network. In *Proceedings of the 6th International Workshop on Climate Informatics: CI*.

Chapter 5
Cognitive Computing, Emotional Intelligence, and Artificial Intelligence in Healthcare

Mohamed Alloghani, Christopher Thron, and Saad Subair

5.1 Introduction

The past few decades have witnessed an extraordinary increase in artificial intelligence, defined as the simulation of intelligent behavior in computing devices. In the healthcare field, artificial intelligence is also gaining popularity due to various factors. To start with, artificial intelligence is seen as an intervention for promoting data-driven decision-making. Decisions that are made with supporting data are likely to produce better outcomes (Bali et al., 2019). Artificial intelligence can enhance diagnosis efficiency through the collection and analysis of patient data. Furthermore, artificial intelligence possesses the potential to enhance cost-effectiveness by minimizing errors and waste.

Cognitive computing refers to the utilization of computerized models to simulate the thought process of a typical human being in situations where solutions might be uncertain or ambiguous. Cognition, which encompasses processes such as language, decision-making, attention, memory, and planning, influences the capability of an employee to perform effectively. One aspect of cognition is emotional intelligence,

M. Alloghani (✉)
Computing and Mathematical Sciences, Liverpool John Moores University, Liverpool, UK

The UAE Artificial Intelligence Office, Prime Minister's Office at the Ministry of Cabinet Affairs and the Future, Dubai, UAE

C. Thron
Department of Science and Mathematics, Texas A&M University-Central Texas, Killeen, TX, USA
e-mail: thron@tamuct.edu

S. Subair
College of Computer Studies, International University of Africa, Khartoum, Sudan
e-mail: ssubair@iua.edu.sd

M. Alloghani et al. (eds.), *Artificial Intelligence for Data Science in Theory and Practice*, Studies in Computational Intelligence 1006,
https://doi.org/10.1007/978-3-030-92245-0_5

which may be defined as the capacity of a person to monitor their own feelings and emotions as well as those of others, discriminate them, and use the information acquired to guide decision-making. Four components of emotional intelligence may be identified: perception of emotions, emotional management, comprehension of emotions, and the facilitation of thought based on emotions. Healthcare professionals should possess emotional intelligence as it enhances their ability to offer person-centered services (Prezerakos, 2018). In settings where patients and families are devastated by illness and other health conditions, emotional intelligence is especially critical for caregivers.

While the fundamental use case of artificial intelligence is the implementation of predefined algorithms to solve a problem through analyzing a number of factors, cognitive computing goes a step further and attempts to mimic human intelligence and wisdom. Unlike artificial intelligence systems that provide set solutions to set problems, cognitive computing offers suggestions for people to take appropriate actions based on their own understanding. Cognitive computing can be viewed as taking the role of an assistant, rather than taking full control of a process as AI does. For example, an AI-backed system for medical diagnosis would make all treatment decisions, while cognitive computing would supplement human diagnosis with its own set of data and analysis, helping to improve decision quality while preserving a human touch in critical processes. As this example shows, cognitive computing thus gives people the power to analyze data more quickly and accurately and make better decisions without having to worry about the wrong decisions taken by the machine learning system (Makadia, 2019).

This chapter provides an overview of the current status of cognitive computing in healthcare, by presenting the results of a systematic literature review. The chapter is organized as follows. Section 5.2 elucidates the methods used in the literature review; Section 5.3 describes the individual studies that form the basis of the review; and Sect. 5.4 discusses themes and major trends that may be discerned in the literature.

5.2 Methods

In this systematized literature review, studies relating to cognitive computing, emotional intelligence, and artificial intelligence are obtained and evaluated. Typical of such studies, the initial step was to establish the eligibility criteria. In line with the objectives of this study, the inclusion criteria included studies focusing on cognitive computing, emotional intelligence, and artificial intelligence in healthcare; studies conducted in English; and articles published within the past 5 years. After determining the eligibility standards, the next step entailed selecting appropriate information sources, which included Scopus, EBSCOhost, MEDLINE, and Cochrane Library. The choice of these databases was premised on the need to identify peer-reviewed and current resources related to the research topic. Additionally, these databases employ diverse tools enabling one to make searches in an easy manner. The search was conducted between January 20 and February 4, 2020.

In addition to reading the abstracts, the author examined each article with a view of establishing credibility and trustworthiness. Critical areas of focus included the literature review, research methodology, data collection and analysis, discussion, and funding sources. Thereafter, the systematic review was done by comparing the results and findings of the studies selected. To execute the search, several search terms were utilized. They included emotional intelligence, cognitive computing, artificial intelligence, and healthcare. Different combinations of these terms, using Boolean operators, were used to search for articles. The output was limited to studies conducted within the previous 5 years and published in English.

The flow diagram in Fig. 5.1 summarizes the search process, from the initial search to the articles finally selected for systematized review. The initial search described above produced 58 articles. An additional ten articles were obtained by searching electronic databases and reference lists of studies. After duplicate titles were removed, 32 articles remained, of which 12 were excluded based on title and abstract review. An additional five references were excluded because of duplication. The remaining 15 articles were examined in detail for eligibility. Content quality, research quality and relevance to cognitive computing, emotional intelligence, and artificial intelligence in healthcare were the primary criterion for eligibility. The final outcome of this selection process was the seven studies that formed the basis for the review.

5.3 Results

Of the seven studies obtained through the selection process, the vast majority were secondary research articles ($n = 6$), while only one was a primary study. The primary study was randomized research in which the authors examined the impact of an AI-based intervention on depression and anxiety. The other six studies entailed examining prior literature to demonstrate the application of AI and cognitive computing in healthcare. Additionally, the articles assessed some of the challenges associated with these emerging technologies. The following paragraphs provide an overview of each study.

The first study selected is a systematic literature review by Behera, Bala, and Dhir, in which the authors explored the emerging role of cognitive computing in healthcare (Behera et al., 2019). The systematic review of literature aimed at examining prior research methods, applications, algorithms, results, and strengths and weaknesses of articles in relation to cognitive computing. According to the findings, cognitive computing possesses the potential of enhancing the quality-of-care delivery.

In the second study, Chen, Li, Hao, Qian, and Humar proposed a cutting-edge cognitive computing system for use in healthcare environments (Chen et al., 2018). The researchers point out that patients in emergency departments are

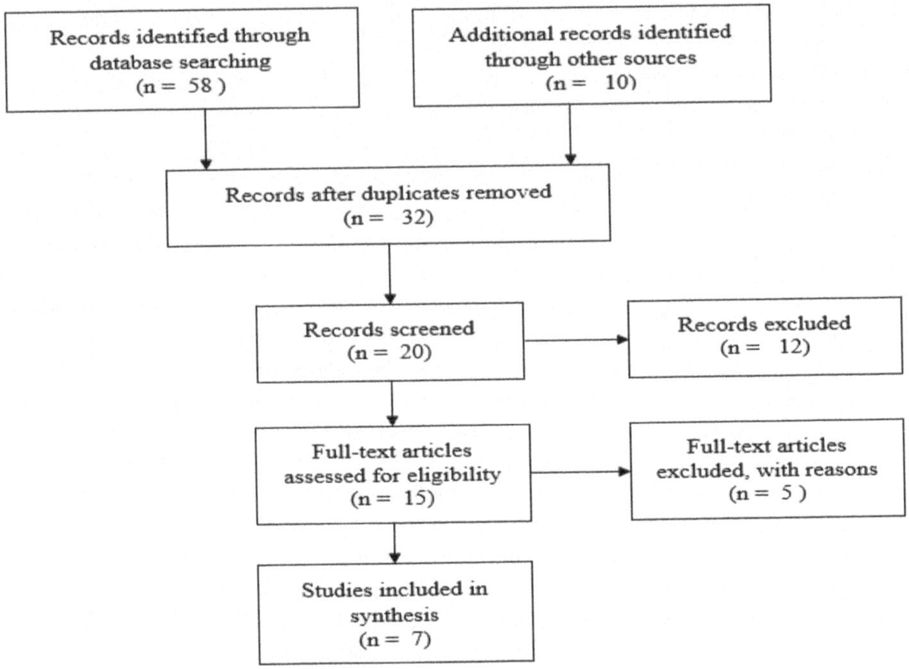

Fig. 5.1 Flow diagram depicting the search process

usually in a delicate health situation, hence the need for quick diagnosis and intervention. Accordingly, the use of smart technologies offers a timely option for examining emergency patients and recommending treatments. The system created had the capacity to monitor and evaluate the physical health of patients by utilizing cognitive computing. Additionally, the researchers incorporated resource allocation capability, so that patients at a higher risk are accorded more resources. After testing the system, the researchers found that it had the capacity to assess patients effectively and efficiently, hence paving the way for care interventions to be implemented. Moreover, the system enhanced the cost-effectiveness of care delivery and improved patients' care experience. Furthermore, the researchers highlighted the potential of using a similar system to examine a patient's emotions during the care provision process (Chen et al., 2018). Besides illustrating the applicability and effectiveness of cognitive computing, the article presents the method of operation of artificial intelligence systems in healthcare.

The study of Davenport and Kalakota explores the applications of artificial intelligence in healthcare and its future prospects (Davenport & Kalakota, 2019). Their article summarizes some of the types of AI in use today in healthcare environments. They include machine learning, natural language processing, and rule-based expert systems. Other applications include robotic process automation and physical robots. One of the challenges associated with the adoption of AI and other technologies in the healthcare environment is the risk of job losses due to automation. There are also ethical implications associated with AI, including the risk to data privacy and confidentiality and the lack of accountability when machines are used to drive decision-making (Davenport & Kalakota, 2019). Despite these concerns, the healthcare industry is likely to see an increase in AI over the coming years.

Liyanage et al. conducted a three-round Delphi study to determine healthcare professionals' perceptions regarding the issues, perceptions, and challenges of artificial intelligence in primary care (Liyanage et al., 2019). The study involved three rounds: the first involved a discussion of 20 primary healthcare professionals on AI; the second focused on the assessment of the information obtained in the initial discussion; and the final round was an online discussion on the findings. According to the findings, healthcare professionals perceive AI as having the potential to improve clinical and administrative processes and decisions. However, the study failed to highlight the effect of AI on the coordination and continuity of care in primary care settings. The participants also stressed the need to promote the ethical utilization of technological innovations.

In the only randomized controlled study selected, Fulmer et al. explored the use of psychological artificial intelligence to relieve symptoms of anxiety and depression (Fulmer et al., 2018). One of the problematic areas of artificial intelligence is whether it has the capability to comprehend the emotions of people and give an appropriate intervention. In this study, an AI chatbot named "Tess" (developed by the X2AI Foundation, x2ai.com) was used to deliver personalized conversations based on the patient's expressed emotions and concerns. A sample of 74 participants was randomized into either the control or intervention group. Based on the results, the authors found a significant reduction in the symptoms of anxiety and depression within the intervention groups as compared to the control group. As a result, they concluded that AI is a cost-effective therapeutic agent (Fulmer et al., 2018). However, they also emphasized that AI cannot replace the trained specialist but can be used to offer patient support. The key strength of this study is that it demonstrated the capability of contemporary AI to read emotions and offer treatment recommendations.

In another study, Loh performed a qualitative review of the literature to illustrate the rise of robots in medicine (Loh, 2018). The author summarized the body of research literature for artificial intelligence in health published in 2017, covering diverse specialties while examining the risks and strengths associated with this emerging innovation. The main finding is that artificial intelligence is effective in diagnosing different conditions and in some cases can be more accurate than humans (e.g., in suicide prediction). The main reason for this is that AI possesses the capability to recognize patterns in large datasets and learn from trends. Additionally, the study highlighted liability and attribution of negligence in cases of medical errors as areas of legal and ethical concern (Loh, 2018). At a time when accountability is expected from healthcare professionals, it becomes difficult to establish liability when machines provided information that supported wrong decisions.

The final study by Marshall et al. assessed cognitive computing and e-science in health sciences research (Marshall et al., 2017). The objective of the authors was to present research frameworks based on AI and evaluate the concepts of e-science and cognitive computing as disruptive factors in the healthcare industry. According to the article, the increasing utilization of artificial intelligence is driven by big data. The healthcare industry today employs many technologies that collect enormous amounts of data from individuals and other sources. This data can be analyzed using different models to augment or amplify the performance of human beings. Some of the models the paper presents include conventional computer support, federated cognition with cognitive agents and machine learning, and semi-autonomous cognitive model based on deep machine learning (Marshall et al., 2017). The study also presents a case study in which artificial intelligence models were used to predict obesity and recommend effective interventions. The overarching view is that AI enhances clinical decision-making resulting in improved patient outcomes.

5.4 Discussion

Although artificial intelligence and cognitive computing are promising interventions, they are still deficient in diverse ways. Firstly, the adoption of technologies in healthcare presents a risk of data security and privacy. Today, healthcare organizations are employing sophisticated technologies to collect patient data to improve service delivery and decision-making. Then collected data is susceptible to cyber-attacks, hence emphasizing the need to implement effective security measures. Secondly, the adoption of artificial intelligence requires a high initial capital investment. Thirdly, organizations might encounter reluctance from staff members when introducing artificial intelligence tools within the workplace. Moreover, the link between the utilization of these tools and the improvement in patient outcomes remains unclear (Bali et al., 2019).

The most evident theme from the reviewed articles is the application of artificial intelligence and cognitive computing diagnosis and treatment recommendations (Behera et al., 2019; Chen et al., 2018; Davenport & Kalakota, 2019; Liyanage et al., 2019; Fulmer et al., 2018; Loh, 2018; Marshall et al., 2017). Today, healthcare workers can rely on rule-based systems and algorithms to diagnose patients. Essentially, each disease or ailment has signs and symptoms. Additionally, there are risk factors that enhance the likelihood of a certain disease occurring. At the same time, there has been a proliferation of information technologies in healthcare environments with the capacity to collect huge amounts of data. Consequently, cognitive computing and AI can be used to analyze patient data to diagnose diseases (Davenport & Kalakota, 2019). Similarly, these emerging technologies can be used to recommend treatment based on factors such as the disease, the nature of the patient, and the available interventions.

AI and cognitive computing are particularly suited to diagnosis and treatment recommendations because they are accurate and efficient. Human beings are limited in terms of their level of precision and discernment and are subject to biases. Accordingly, healthcare workers are prone to unintentional errors. During diagnosis, a health professional might fail to notice something, hence making the wrong observation. As a result, AI and cognitive computing are more accurate in medical diagnosis as compared to human beings in some circumstances (Behera et al., 2019; Chen et al., 2018; Davenport & Kalakota, 2019; Liyanage et al., 2019; Fulmer et al., 2018; Loh, 2018). For example, AI is more accurate in predicting suicide than people. In radiology, AI is outperforming professionals in diagnosing malignant tumors (Chen et al., 2018). Furthermore, AI is effective in predicting obesity and other lifestyle diseases (Marshall et al., 2017). The main finding is that AI and cognitive computing are able to examine data and make an accurate and timely diagnosis. Therefore, they are an important addition in the contemporary healthcare industry as they possess the capability to enhance the safety and quality of care delivery.

The patient-centered paradigm has been championed as an approach that can enhance the effectiveness of the healthcare system. Essentially, healthcare providers are required to offer services in a manner that aligns with the needs, preferences, and wants of their clients. To this end, artificial intelligence provides a means for enhancing patient engagement (Davenport & Kalakota, 2019). Patient engagement allows one to understand the challenges a patient is facing, including socioeconomic issues, and provide care that is person-centered. According to Davenport and Kalakota, AI-based capabilities are effective in contextualizing and personalizing care (Davenport & Kalakota, 2019). Tools such as machine learning allow healthcare workers to understand their patients deeply, hence enabling them to offer targeted communications and interventions.

In addition to mimicking human cognitive abilities, current AI tools have some capacity to read, understand, and respond to emotions. One of the features that make health-related careers different from other professions is that the client is a patient or family suffering from health-related issues. Therefore, healthcare workers are required to exhibit compassion and other people-oriented skills. Emotional intelligence, which is characterized by self-awareness, motivation, empathy, self-regulation, and social skills (Bali et al., 2019), is an essential competency. Some of the studies examined showed the possibility of using AI to offer emotionally satisfactory care to patients with symptoms of depression and anxiety (Loh, 2018; Fulmer et al., 2018). One particularly prominent example is the Tess chatbot, which has been accessed online by almost 20 million people and has demonstrated positive mental health outcomes in several independent research studies (x2ai.com/outcomes). However, AI has not fully reached a point where it can match human emotional capacity, and, therefore, such tools should only complement the service of human professionals.

Beyond the bedside, artificial intelligence and cognitive computing can help in streamlining administrative tasks (Davenport & Kalakota, 2019; Liyanage et al., 2019). Some of the tasks healthcare workers and other employees in the healthcare sector perform include preparing reports, maintaining patient records, updating health records, managing expenses, and tacking medical supplies. Artificial intelligence and cognitive computing provide ways of assessing records to identify trends and insights. Using the information obtained from these systems, healthcare workers can make appropriate changes. For example, an algorithm can be developed to examine internal processes to establish factors contributing to the inability of a facility to offer timely services. With this information, the leadership can implement a quality improvement program to address problem areas.

The assessment of the studies obtained also highlighted challenges associated with cognitive computing and artificial intelligence. The first one relates to automation and its potential of contributing to job losses (Davenport & Kalakota, 2019; Marshall et al., 2017). Evidently, technological interventions are able to make accurate and timely decisions. In some situations, they are more accurate as compared to human beings. Additionally, the possibility of human error is eliminated when these emerging technologies are utilized. Therefore, it seems logical for the healthcare industry to continue investing in technology both as a way of improving patient outcomes and lowering human resource expenses. The problem with this approach is that it would lead to job losses. However, this risk is minimal mainly because healthcare workers do more than just assessing data in diagnosing a patient. Other roles such as engaging patients and their families can only be performed effectively by human beings.

Another concern associated with AI and cognitive computing is data confidentiality and privacy (Fulmer et al., 2018; Marshall et al., 2017; Colorafi & Bailey, 2016; Behera et al., 2019). Over the recent past, the healthcare industry has seen the enactment of laws and regulations aimed at protecting patient information. In addition to ensuring that data is secured, the law expects health organizations to use patient data for the intended purposes only. Consequently, the collection of patient data and mining it to improve care delivery or enhance internal administrative processes presents legal issues. In the USA, the Health Insurance Portability and Accountability Act (HIPAA) is key legislation that offers data security and privacy provisions for safeguarding medical information (Behera et al., 2019). Based on this law, organizations are required to implement physical, administrative, and technical safeguards to guarantee the security and privacy of patient information (Colorafi & Bailey, 2016). Likewise, the adoption of AI and cognitive computing should incorporate these safeguards.

The final concern relates to legal liability when AI and cognitive computing are used to influence decision-making (Loh, 2018). The application of AI in healthcare largely entails analyzing patient data to diagnose diseases and make treatment recommendations. Although AI is perceived as being more accurate as compared to human beings, it often makes the wrong diagnosis resulting in the implementation of the wrong intervention resulting in adverse consequences. In such situations, it would seem illogical holding to account the healthcare worker for the outcome of those decisions.

5.5 Conclusions

This systematized literature review examined cognitive computing, emotional intelligence, and artificial intelligence in healthcare. Cognitive computing entails creating computerized models to imitate the thought process of human beings. Artificial intelligence is a similar term as it refers to the simulation of human intelligence in computers. Emotional intelligence encompasses being aware of and managing one's emotions, noticing and comprehending the emotions of others, and managing relations with others. Emotional intelligence is necessary for healthcare workers as they work with people on a daily basis. Similarly, cognitive computing and artificial intelligence are emerging technologies that can be used to augment professionals' decision-making.

A literature search was done, and seven studies were selected for the review. An analysis of those studies highlighted several themes. To begin with, artificial intelligence and cognitive computing enhance disease diagnosis and treatment recommendations. Generally, these technologies examine patient data to determine what might be wrong and recommend the best intervention for optimal outcomes. In addition, cognitive computing and artificial intelligence enhance patient engagement and can help in streamlining administrative tasks. They can also help in reading patients' emotions and respond to them. Lastly, the study found several challenges associated with AI and cognitive computing. The most common issue was the risk to the privacy and security of data. Moreover, these emerging technologies increase automation, which is associated with job losses. Furthermore, the issue of legal liability in case of a medical error is apparent when AI and other technologies are used in decision-making. Overall, despite the challenges, AI and cognitive computing are likely to revolutionize healthcare in the near future.

References

Bali, J., Garg, R., & Bali, R. T. (2019). Artificial intelligence (AI) in healthcare and biomedical research: Why a strong computational/AI bioethics framework is required? *Indian Journal of Ophthalmology, 67*(1), 3–6. https://doi.org/10.4103/ijo.IJO_1292_18

Behera, R. K., Bala, P. K., & Dhir, A. (2019). The emerging role of cognitive computing in healthcare: A systematic literature review. *International Journal of Medical Informatics, 129*, 154–166. https://doi.org/10.1016/j.ijmedinf.2019.04.024

Chen, M., Li, W., Hao, Y., Qian, Y., & Humar, I. (2018). Edge cognitive computing based smart healthcare system. *Future Generation Computer Systems, 86*, 403–411.

Colorafi, K., & Bailey, B. (2016). It's time for innovation in the health insurance portability and accountability act (HIPAA). *JMIR Medical Informatics, 4*(4), e34. https://doi.org/10.2196/medinform.6372

Davenport, T., & Kalakota, R. (2019). The potential for artificial intelligence in healthcare. *Future Healthcare Journal, 6*(2), 94–98. https://doi.org/10.7861/futurehosp.6-2-94

Fulmer, R., Joerin, A., Gentile, B., Lakerink, L., & Rauws, M. (2018). Using psychological artificial intelligence (tess) to relieve symptoms of depression and anxiety: Randomized controlled trial. *JMIR Mental Health, 5*(4), e64. https://doi.org/10.2196/mental.9782

Liyanage, H., Liaw, S. T., Jonnagaddala, J., Schreiber, R., Kuziemsky, C., Terry, A. L., & de Lusignan, S. (2019). Artificial intelligence in primary health care: Perceptions, issues, and challenges. *Yearbook of Medical Informatics, 28*(01), 041–046.

Loh, E. (2018). Medicine and the rise of the robots: A qualitative review of recent advances of artificial intelligence in health. *BMJ Leader, 2018*. https://doi.org/10.1136/leader-2018-000071

Makadia, M. (2019). *What is cognitive computing? How are enterprises benefitting from cognitive technology?*https://towardsdatascience.com/what-is-cognitive-computing-how-are-enterprises-benefitting-from-cognitive-technology-6441d0c9067b

Marshall, T., Champagne-Langabeer, T., Castelli, D., & Hoelscher, D. (2017). Cognitive computing and eScience in health and life science research: Artificial intelligence and obesity intervention programs. *Health Information Science and Systems, 5*(1), 13. https://doi.org/10.1007/s13755-017-0030-0

Prezerakos, P. E. (2018). Nurse managers' emotional intelligence and effective leadership: A review of the current evidence. *The Open Nursing Journal, 12*, 86–92. https://doi.org/10.2174/1874434601812010086

Chapter 6
A Systematic Review on Application of Data Mining Techniques in Healthcare Analytics and Data-Driven Decisions

Mohamed Alloghani, Saad Subair, and Christopher Thron

6.1 Introduction

Regardless of the economic or sociopolitical conditions prevailing in any country, the healthcare industry invariably finds avenues in which to thrive (Brandao de Souza, 2009). Even in the least-developed countries, governments go to great lengths to outsource medical experts from other countries, and many people (especially the wealthy) continue to pursue treatment in different parts of the world, particularly in certain countries (such as India and the United States), which have developed a reputation for higher-quality healthcare (Thakur et al., 2012).

In addition, countries worldwide are migrating toward electronic health (e-health), and medical data is increasingly formatted as electronic medical records. This migration involves standardizing the medical data and ensuring that aggregated patient clinical information can be accessed instantly, so that the information retrieved can support evidence-based practice moving forward (Priyanka & Kulennavar, 2014). Although these developments have procured benefits, they also brought serious challenges, such as exponentially growing costs, inefficient services,

M. Alloghani (✉)
Computing and Mathematical Sciences, Liverpool John Moores University, Liverpool, UK

The UAE Artificial Intelligence Office, Prime Minister's Office at the Ministry of Cabinet Affairs and the Future, Dubai, UAE

S. Subair
College of Computer Studies, International University of Africa, Khartoum, Sudan
e-mail: ssubair@iua.edu.sd

C. Thron
Department of Science and Mathematics, Texas A&M University-Central Texas, Killeen, TX, USA
e-mail: thron@tamuct.edu

M. Alloghani et al. (eds.), *Artificial Intelligence for Data Science in Theory and Practice*, Studies in Computational Intelligence 1006,
https://doi.org/10.1007/978-3-030-92245-0_6

increased readmission rates for communicable diseases, and increasing difficulty of handling large volumes of data (Priyanka & Kulennavar, 2014).

Of the many solutions that have been proposed to handle these challenges, data mining is the most promising. Unlike conventional data analysis which relies on statistical tests, data mining uses more computationally intensive techniques such as text mining, association mining, supervised machining learning, and unsupervised machine learning which can provide more detailed information (Shah & Tenenbaum, 2012). Data mining techniques have been used across different fields and industries for over 40 years. However, the healthcare industry poses an especially difficult challenge because of the differences in the nature of the data collated as well as the standards (Garrison, 2013). The growing volumes of healthcare data are unstructured and lack concise semantics. Additional challenges include difficulty in maintaining patient privacy, adherence to legal requirements, and imposition of restrictions that hinder knowledge discovery from the data (Kuo et al., 2014). Nonetheless, experts are optimistic that data mining in conjunction with information technology will revolutionize the healthcare industry and integration and inclusion of big data and other related *Industry 4.0* concepts will improve service provision, diagnosis, and treatment (Foster, 2014; Feldman et al., 2012).

Several previous studies have examined the trends and the prospective applications of data mining in healthcare analytics. Most of these focus on specific applications or on selected algorithms and how such algorithms can be implemented to solve problems associated with specific conditions and diseases (Kuo et al., 2011). Of particular interest is the application of data mining techniques in healthcare analytics. Data analytics in general refers to quantitative and qualitative techniques deployed to synthesize data and represent the synthesized data to provide significant insights. Such tools provide information used for planning as well as management of a variety of real-life applications. For example, data analytics has been used to identify fraudulent readmission cases (Youngstrom, 2012). In general, the application or integration of analytics, including data mining and machine learning algorithms, is assisting doctors and nurses in the implementation of more accurate diagnosis, treatment, and furthering prospects for predictive medicine (Youngstrom, 2012). The applications of data analytics in healthcare can be viewed from either an applied or theoretical perspective. For example, some previous application-oriented studies have focused on the application of machine learning in either mental health or pharmacovigilance, while theoretical oriented studies have dealt with methods and challenges of data mining for health applications. The body of literature on health analytics has been growing, and this systematic review summarizes and organizes peer-reviewed published papers that focus on applied and/or theoretical perspectives.

6.1.1 Motivation and Scope

The number of papers published on healthcare and data mining has grown exponentially over the past 10 years. Even though several published papers have reviewed healthcare analytics and the role of data mining, very few have employed the systematic review approach. Table 6.1 summarizes four systematic review papers that have been published on the application of data mining health analytics. Of these four, one article (Youssef, 2014) focused on the papers that were published between 1977 and 2011, and the reviewers concluded that the number of published papers had risen from 5 in the 1990s to over 726 publications in 2012. The paper did not specify the specific data mining techniques that were discussed in these publications. The review period for article (Niaksu et al., 2014) was only 2 years; nonetheless the researchers identified 226 relevant articles from PubMed, Scopus, and Web of Science and managed to isolate 92 full-text relevant papers. However, it should be noted that the scope of the review was limited to mobile data mining applications. Article (Fallah & Niakan Kalhori, 2017) reviews papers covering both applied and theoretical perspective of data mining applications in healthcare analytics. The findings and conclusions of the review suggest that most healthcare analytics are mainly used for administrative and clinical decision-making and that electronic health records are the leading source of medical data. However, the review suggests that prescriptive analytics are lacking and that real-time integration of expert knowledge in clinical decision-making is not receiving sufficient attention. Finally, reference (Islam et al., 2018) focused on descriptive studies of the applications of big data analytics in both medicine and healthcare. The search scope excluded all numerical or quantitative studies that involved either clinical trials or practical implementation of analytical techniques on patient data.

This chapter extends the previous literature by systematically reviewing literature on data mining techniques and their applicability in healthcare regardless of the condition. The scope includes both theoretical and analytical articles between 2005 and 2018 and covers the impacts of analytics on all areas of healthcare, as well as the data mining algorithms used (and their variants). As an approach, the systematic review included papers regarded as literature review or simply reviews as part of the papers investigated.

Table 6.1 Characteristics of some of the existing systematic review papers on healthcare analytics and data mining

Review article	Scope	Review period	Papers reviewed
Youssef (2014)	Data mining applications in healthcare	1997–2012	NA
Niaksu et al. (2014)	Data mining applications in patient-centered mobile-based information systems	2014–2016	92
Fallah and Niakan Kalhori (2017)	Application and theoretical perspective of data mining	2005–2016	117
Islam et al. (2018)	Concurrence of big data analytics and healthcare	2013–2018	58

6.2 Methodology

The literature search was conducted in 2019. The following sections describe first the literature search and article selection procedures, then the methods used for quality assessment and analysis of article contents.

6.2.1 Literature Search and Article Selection

Figure 6.1 shows the four phases of PRISMA framework that was used to guide the article selection process (Singh, 2013). These phases are described in more details as follows.

The first PRISMA phase (identification) involves a series of Internet searches for potential resources. The search queries were primarily implemented on the ProQuest Central database, with the queries in subsequent phases based on the modification of previous queries, including specification on the type of the documents, subject areas, the range of the year of publication, and whether the chosen article is peer-reviewed or full text. The search keywords were all related to applications of data mining

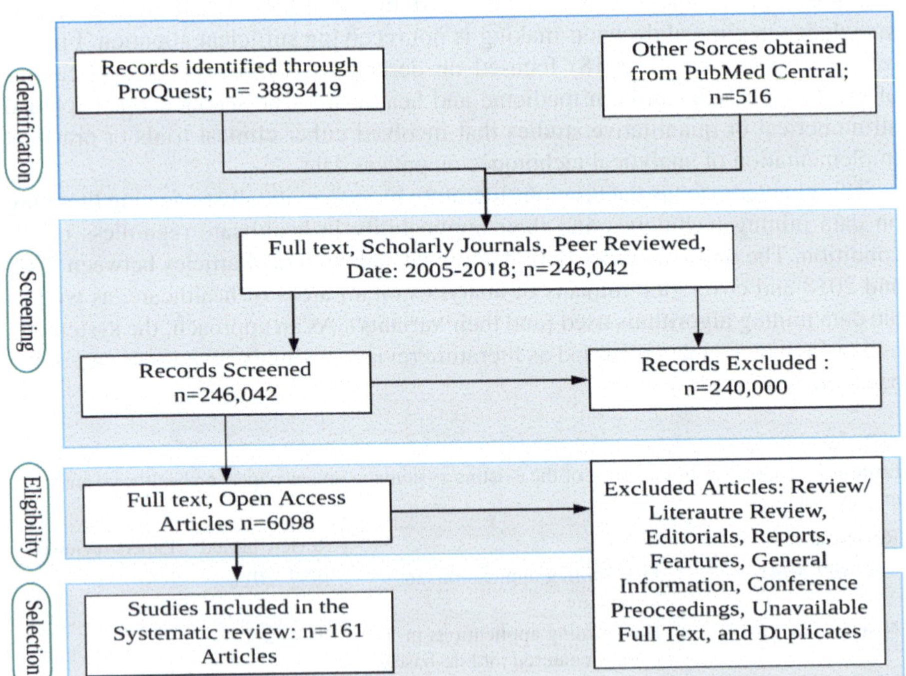

Fig. 6.1 Preferred reporting items for PRISMA framework flow, demonstrating the literature review process

in healthcare, and different variants were used. The search focused on the 2005–2018 period, but unlike the review paper (Fallah & Niakan Kalhori, 2017) which covers a similar period, this systematic review excluded papers that discussed data mining in other fields besides health. The primary filters used in the preliminary searches included language setting, database specification, and publication period. The first search resulted in 389,914 search hits, and specifications of full-text and peer-reviewed filters reduced the number to 247,469. Additional strategies used in reducing the number of hits included narrowing down to scholarly articles, which resulted in 246,042 "qualified" articles.

The second phase (screening) excluded articles that were published as reviews, editorials, working papers, conference papers (except those presented at IEEE conferences), reports, features, and general information. Duplicates and articles without open-access full text were also excluded. An attempt to search for articles with "healthcare" as part of the publication and "data mining" as designated search terms yielded 56 additional articles, which when added to the screened articles resulted in a total of 6098 articles that were passed to the "eligibility" phase.

In the third phase (eligibility), the title, abstract, and keywords of the remaining articles were visually examined to identify papers discussing the applicability of data mining in healthcare, especially its support of or influence on healthcare data analytics. The result was a total of 161 articles that formed the basis of this review.

6.2.2 *Quality Assessment and Processing Steps*

The 161 articles were categorized using a hierarchical classification as follows. Within each of these groups, the articles were further classified according to objectives, data types, algorithms used, and applications. For example, Fig. 6.2 shows the further classification of studies of applications of data mining in healthcare. The figure indicates that these papers focused on predictive and prescriptive objectives; employed data from various sources, including sensors, biometrics, EMR, and transactional data (data recorded from digital transactions), as well as e-Mobile data; employed classification and clustering algorithms, as well as association rules mining and anomaly detection; and represented several key healthcare areas (only the top eight are shown in the figure).

The quality of papers dealing with applications of data analytics was assessed using the JBI Critical Appraisal Tool (the Joanna Briggs Institute (JBI) is an international, not-for-profit research and evidence-based practice healthcare center based in the University of Adelaide, South Australia.), while the Critical Appraisal Skills Programme (CASP) was used to appraise theoretical papers (Moola et al., 2020). The quality assessment process involved checking for concisely stated research objectives with sample population selection and inclusion criteria, comprehensiveness of the description of the variables, the clarity on the source of the data (hospital or survey), as well as the format of the data. The other elements considered included validity and reliability of the data collation instrument, measures of the

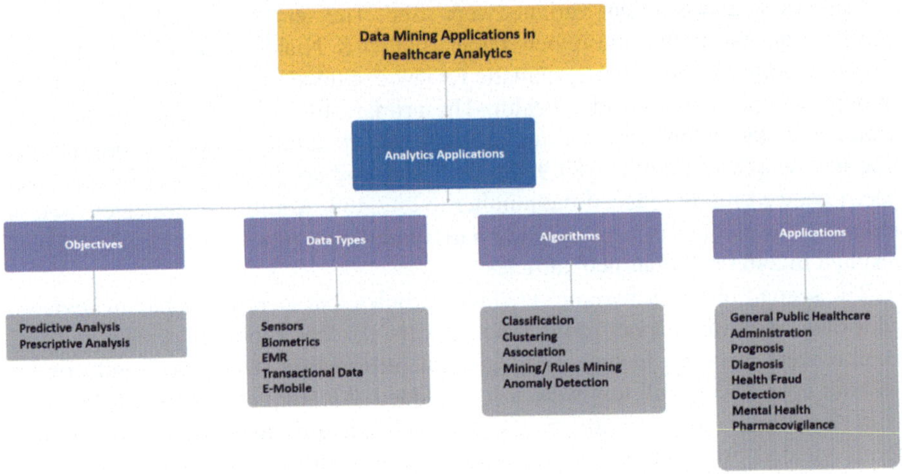

Fig. 6.2 Further classification of analytics papers by objective, data types, algorithms used, and applications

health outcomes, and inclusion of the analytical tool used in the study. Besides these factors, the researcher considered the clarity of stated aims, the appropriateness of used qualitative methods as well as the design of the study, clarity of the findings, and the general contribution of such papers to healthcare analytics.

6.3 Results

The findings of the review are presented based on the number of articles published per year and distribution per journal, including a comprehensive analysis of the quality of each paper.

This section is organized as follows: Sect. 6.3.1 is about the distribution of papers based on year of publication; Sect. 6.3.2 talks about the distribution of papers based on publishing journals; Sect. 6.3.3 presents the healthcare analytics types based on literature search results; and Sect. 6.3.4 reviews the application of analytics in healthcare; while Sect. 6.3.5 presents the theoretical studies.

6.3.1 Distribution of Papers Based on Year of Publication

Figure 6.3 shows a bar chart for the year of publication for the reviewed papers. All papers used for data mining as well as analysis of big data and the trend show a generally growing interest among stakeholders in the healthcare sector,

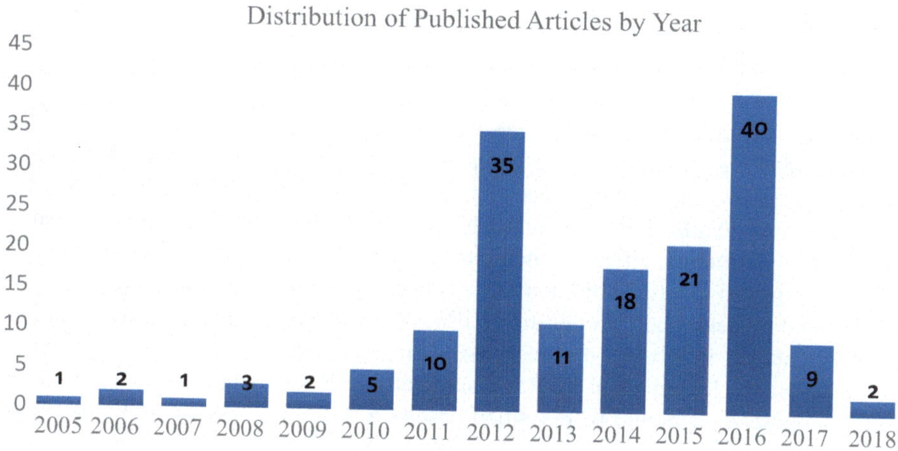

Fig. 6.3 Year of publication for the 161 reviewed papers

Table 6.2 Top 10 journals based on the literature search results

Journals articles	Articles published
Knowledge and Information Systems	18
Plos One	10
BMC Medical Informatics and Decision Making	5
Biomed Research International	4
Information Systems Frontiers	4
Journal of the Association for Information Systems	4
Multimedia Tools and Applications	3
Annals of Operations Research	2
BMC Bioinformatics	2
Business and Information Systems	2

peaking with 40 references in 2016. Stakeholders include governmental agencies, non-governmental agencies, healthcare providers, and academicians in the sector, as well as other parties. The trend is an indication that the concerned parties are likely to continue applying analytics for the improvement of the healthcare sector.

6.3.2 *Distribution of Papers Based on Publishing Journals*

In total, all the papers used in the current study were distributed in 70 different scientific journals. Table 6.2 presents the ten journals that published the greatest number of articles used in the current analysis, which together account for 33% of the articles analyzed.

6.3.3 Healthcare Analytics Types Based on Literature Search Results

Due to the differences in study areas, consequently datasets which are used in analytics also had widely varying characteristics. Different categories of datasets are used: human-generated datasets such as EMR (electronic medical records) (the EMR contains details about a patient's health profile, for instance, treatment plans, prescription diagnoses, allergies, medicines, lab tests, and immunizations) and EHR (electronic health records) confidential details of a patient's health history, including all past and present medical conditions, illnesses, and treatments. The classification of these datasets was based on the classification scheme developed by Raghupathi and Raghunath (Karimi et al., 2015; Raghupathi & Raghupathi, 2014). According to the scheme, data should be classified based on its nature, source, and method of data collection.

6.3.4 Application of Analytics in Healthcare

Surveyed studies were concerned with the application of a variety of analytical techniques to healthcare related to data mining. Many of these are related to applications related to the prediction, diagnosis, and prognosis of various medical conditions. In this regime, a wide variety of applications are treated, including such diverse topics as warehousing of clinical data for disease support, the analysis of risk of conditions such as cardiovascular disease, data mining for anxiety disorders, monitoring of wearables, analysis of ICU mortality rates, asthma attack prediction, and so on (Yeh et al., 2009; Nguyen et al., 2015; Zhou et al., 2010; Erinjeri et al., 2009; Praveenkumar et al., 2014; Panagiotakopoulos et al., 2010; Lavraˇc et al., 2007; Harpaz et al., 2012; Luo & Grams, 2012; Alam & Ben Hamida, 2015; Alaniz et al., 2017; Alickovic & Subasi, 2016; Aref-Eshghi et al., 2017; Baig et al., 2017; Matsoukas et al., 2015; Berikol et al., 2016; Bhatia & Sood, 2016; Kuo et al., 2015; Chen et al., 2015a, b, 2016; Chen & Zhong, 2012; Cheng et al., 2015, 2016, 2017; Chiang & Pao, 2016; Choi et al., 2016; Cicirelli et al., 2016; Diz et al., 2016; Dobalian et al., 2012; Oh & Teege, 2011; Fan et al., 2016; Faria et al., 2015; Faust et al., 2012). Some studies also supported administrative functions, emphasizing the importance of administrative support on health outcome efficacy.

In the following subsection, we briefly discuss some prominent active application areas.

6.3.4.1 Cardiovascular Diseases

Despite the failure of a literature search to narrow down search results based on diseases, some of the papers focused on specific disease, notably cardiovascular ones and diabetes (Figueroa & Flores, 2016; Forsvik et al., 2017; Gambhir et al., 2016; Grams, 2012a; Guédon et al., 2016). The commonly addressed cardiovascular conditions are related to coronary artery surgery and myocardial infarction. The studies suggested that age and sex or gender were the most obvious predisposing factors although factors such as smoking habit and medical history also emerged as some of the most prominent factors. The most notable diagnostic application of data mining is in the study of coronary disease since it is easier to identify patterns from demographical and medical history records.

Based on the findings presented in the analyzed papers, it emerged that traditional data analysis techniques such as probabilistic forecast are highly uncertain and have considerably higher misclassification margins (Guo et al., 2015, 2016; Hristovski et al., 2016).

6.3.4.2 Diabetes

The diabetic application of data mining has been driven by the fact that diabetes is a lifestyle disease that continues to a major cause of mortality in developed countries (Hsieh et al., 2010, 2012; Huang et al., 2016; Khennak & Drias, 2017; Koutkias & Jaunt, 2016; Koyuncugil & Ozgulbas, 2010a,b). The research on the application of data mining and, in particular, the application of machine learning algorithms in identification and prediction of diabetes pathways has been around for a while, and most of them revolve around the importance of medical information to identify complications related to diabetes (Kozat et al., 2009; Kupusinac et al., 2016; Lalos et al., 2016; León et al., 2016; Lin et al., 2015, 2016a, b; Lin & Hsieh, 2015; López et al., 2017). Algorithms such as K-nearest neighbor, support vector machine, and random forests have been applied in different studies to the predictive occurrence of different medical conditions. Using predictors and predisposing conditions such as smoking habits, age, body mass index, and hypertension, different studies have demonstrated that machine learning can leverage the information to predict or diagnose the condition of the patient (López et al., 2016; López-Nava et al., 2017; Lv et al., 2016; Magalhães et al., 2016; Melillo et al., 2015; Mezghani et al., 2015; Monteiro et al., 2017; Mudumbai et al., 2017; Mueen et al., 2010; Neves et al., 2015; Ng et al., 2011; Niwas et al., 2015; Nowaková et al., 2017; Oliveira et al., 2017). However, regardless of the specific condition applications, machine learning and artificial intelligence algorithms have been applied.

6.3.4.3 Cancer Diagnosis and Prediction Application

Cancer remains one of the lethal diseases in contemporary society because its treatment is elusive, while it continues to affect millions of people. Machine learning and deep learning technologies have been applied to different aspects of cancer analytics (Osborne et al., 2017; Palacio et al., 2010; Pérez et al., 2015; Preve, 2011; Pustisek, 2017; Rafferty et al., 2015; Ramanan et al., 2016; Ramos et al.,

2016). Despite these advances, cancer treatment remains a mystery although data mining and other analytical techniques have made possible for doctors to device chemotherapy and other survival treatments. Algorithms such as the support vector machine and decision trees have been used to predict lung cancer among patients (Rios-Alvarado et al., 2015).

6.3.4.4 Healthcare Administration

One of the emerging applications of data mining is in the field of healthcare administration, which includes the use of data mining to manage waiting times, reduces time to admission, facilitates doctor or nurse contact with patience, and facilitates the overall patient management (Roris et al., 2016; Roy Chowdhury et al., 2009; Sareen et al., 2016; Sayyad Shirabad et al., 2012; Shi et al., 2016; Suciu et al., 2015). Some studies have been conducted on data warehousing and the influence of cloud computing on healthcare analytics (Sufi et al., 2011; Ting et al., 2011; Topan et al., 2016; Triantafyllopoulos et al., 2016; Tsai & Yu, 2016; Tsai et al., 2016; Übeyli & Dodu, 2010; Vanopstal et al., 2011; Verma et al., 2016; Vilhena et al., 2017). However, most of the papers dealing with healthcare administration have focused on the application of machine learning in understanding and minimizing the cost of healthcare and addressing the widespread issue of health quality. Additionally, data mining techniques have been applied to different aspects of patient management studies, including exploration for different management technologies (Villarreal et al., 2015, 2016; Wan & H., 2006; Wanderer et al., 2016; Wang et al., 2015, 2016; Yang et al., 2016a,b).

6.3.4.5 Prognosis and Diagnosis

Several papers discuss the implementation of data mining in healthcare is to improve prognosis/diagnosis and contribute to personalized medicine (Yao et al., 2015; Yu et al., 2016; Yue et al., 2016; Zhang et al., 2016; Ajami & Ketabi, 2012; Aljarullah & El-Masri, 2013; Arif et al., 2014; Baig & Gholamhosseini, 2013; Ben-Assuli et al., 2014). Most of these applications are highly data intensive. The review also revealed that models such as regression trees and their boosted versions can perform better in predicting maladies than others, although this ultimately depends on the available data. Currently ensemble algorithms are trending, and most health analytics applications will have to adopt such algorithms to improve their efficiency and accuracy.

6.3.4.6 Pharmacovigilance

Considering the application of data mining and machine learning algorithms, most of them focus on posttreatment reactions and responses to given drugs and not necessarily from the perspective costs and steps toward ensuring patient safety (Chattopadhyay & Acharya, 2012; Chattopadhyay et al., 2012; Chen et al., 2012, 2014, 2015c; Cho et al., 2015; Chou et al., 2012; De et al., 2015; Deng et al., 2012; Grams, 2012b, c). Most of the pharmacovigilance articles utilize social network data, including the reporting of adverse drug reactions and anticancer agents' data (Gurupur et al., 2012; Hsiao et al., 2012; Hsu & Pan, 2013). Most of these studies deploy advanced probabilistic methods, such as Bayes geometric mean (Huang et al., 2012a, b, 2013; Issac Niwas et al., 2012; Karabulut & Ibrikci, 2014; Karla & Gurupur, 2013).

6.3.5 *Theoretical Studies*

Of the 161 reviewed papers, only 16 papers were considered as focusing on the theoretical perspective of the application of data mining in healthcare analytics (Keltch et al., 2014; Keramidas et al., 2012; Khan et al., 2014; Kohlmann et al., 2014; Korkmaz & Poyraz, 2014; Koyuncugil & Ozgulbas, 2012; Latif et al., 2014; Lee et al., 2013, 2014; Li et al., 2012; Luo, 2013, 2014; Mehmood et al., 2014; Pollettini et al., 2012; Rafe & Hajvali, 2014; Don et al., 2013; Sahin & Celikkan, 2012). However, it should be noted that the search and inclusion criteria eliminated review articles, and as such theoretical papers were based on literature exempted from the study. In general, the theoretical papers discussed the role of healthcare analytics in disease control, data quality control, policies in healthcare, and patient privacy.

6.4 Conclusion

With about 160 papers reviewed, 90% of the papers dealt with applications of data mining (including machine learning) in analytics, while 10% were based on conceptual as well as qualitative aspects.

In analytics, data mining algorithms yield vastly improved accuracy and reliability over traditional data analysis techniques such as probabilistic forecasting. On the administrative side, data mining has been successful in reducing waiting times and time to admission. These algorithms also help in prevention of information loss and are useful in collecting and preparing data for research activities. As far as the decision support side is concerned, machine learning techniques such as support vector machines and decision trees have brought significant improvements in prognosis and diagnosis for a number of diseases (including cancer) and have contributed to personalized medicine. These algorithms also give valuable information on posttreatment reactions and responses to drugs.

Theoretical papers in the recent literature discuss a variety of issues such as the role of healthcare analytics in disease control, data quality control, policies in healthcare, and patient privacy.

Given the rate of growth of data science and its assimilation in the health sector, stakeholders are under pressure to integrate data mining algorithms into current healthcare systems.

References

Ajami, S., & Ketabi, S. (2012). Performance evaluation of medical records departments by analytical hierarchy process (AHP) approach in the selected hospitals in Isfahan. *Journal of Medical Systems, 36*(3), 1165–1171.

Alam, M. M., & Ben Hamida, E. (2015). Strategies for optimal MAC parameters tuning in IEEE 802.15.6 wearable wireless sensor networks. *Journal of Medical Systems, 39*(9), 1–16.

Alaniz, H. O., Abdullah, A. H., & Qureshi, K. N. (2017). A critical review for developing accurate and dynamic predictive models using machine learning methods in medicine and health care. *Journal of Medical Systems, 41*(4), 1–10.

Alickovic, E., & Subasi, A. (2016). Medical decision support system for diagnosis of heart arrhythmia using DWT and random forests classifier. *Journal of Medical Systems, 40*(4), 1–12.

Aljarullah, A., & El-Masri, S. (2013). A novel system architecture for the national integration of electronic health records: A semi-centralized approach. *Journal of Medical Systems, 37*(4), 1–9953.

Aref-Eshghi, E., Oake, J., Godwin, M., Aubrey-Bassler, K., Duke, P., Mahdavian, M., & Asghari, S. (2017). Identification of dyslipidemic patients attending primary care clinics using electronic medical record (EMR) data from the Canadian Primary Care Sentinel Surveillance Network (CPCSSN) database. *Journal of Medical Systems, 41*(3), 1–7.

Arif, M., Bilal, M., Kattan, A., & Ahamed, S. I. (2014). Better physical activity classification using smartphone acceleration sensor. *Journal of Medical Systems, 38*(9), 1–95.

Baig, M. M., & Gholamhosseini, H. (2013). Smart health monitoring systems: An overview of design and modeling. *Journal of Medical Systems, 37*(2), 1–9898.

Baig, M. M., Gholam Hossein, H., Moqeem, A. A., Mirza, F., & Lindén, M. (2017). A systematic review of wearable patient monitoring systems - current challenges and opportunities for clinical adoption. *Journal of Medical Systems, 41*(7), 1–9.

Ben-Assuli, O., Shabtai, I., Leshno, M., & Hill, S. (2014). EHR in emergency rooms: Exploring the effect of key information components on main complaints. *Journal of Medical Systems, 38*(4), 1–36.

Berikol, G., Yildiz, O., & Özcan, I. T. (2016). Diagnosis of acute coronary syndrome with a support vector machine. *Journal of Medical Systems, 40*(4), 1–8.

Bhatia, M., & Sood, S. K. (2016). Temporal informative analysis in smart-ICU monitoring: M-HealthCare perspective. *Journal of Medical Systems, 40*(8), 1–15.

Brandao de Souza, L. (2009). Trends and approaches in lean healthcare. *Leadership in Health Services, 22*(2), 121–139.

Chattopadhyay, S., & Acharya, U. R. (2012). A novel mathematical approach to diagnose premenstrual syndrome. *Journal of Medical Systems, 36*(4), 2177–2186.

Chattopadhyay, S., Kaur, P., Rabhi, F., & Acharya, U. R. (2012). Neural network approaches to grade adult depression. *Journal of Medical Systems, 36*(5), 2803–2815.

Chen, T., & Zhong, S. (2012). Emergency access authorization for personally controlled online health care data. *Journal of Medical Systems, 36*(1), 291–300.

Chen, T., Chung, Y., & Lin, F. Y. S. (2012). Deployment of secure mobile agents for medical information systems. *Journal of Medical Systems, 36*(4), 2493–2503.

Chen, P., Chai, J., Zhang, L., & Wang, D. (2014). Development and application of a Chinese webpage suicide information mining system (SIMS). *Journal of Medical Systems, 38*(11), 1–88.

Chen, D., Chen, Y., Chen, L., Hsu, M., & Chiang, K. (2015a). A machine learning method for power prediction on the mobile devices. *Journal of Medical Systems, 39*(10), 1–11.

Chen, L., Lin, Z., & Chang, J. (2015b). FIR: An effective scheme for extracting useful metadata from social media. *Journal of Medical Systems, 39*(11), 1–14.

Chen, L., Zhang, X., & Wang, H. (2015c). An obstructive sleep apnea detection approach using kernel density classification based on single-lead electrocardiogram. *Journal of Medical Systems, 39*(5), 1–11.

Chen, D., Wang, H., Sheng, L., Hueman, M. T., Henson, D. E., Schwartz, A. M., & Patel, J. A. (2016). An algorithm for creating prognostic systems for cancer. *Journal of Medical Systems, 40*(7), 1–10.

Cheng, C., Lu, C., Hsieh, T., Lin, Y., Taur, J., & Chen, Y. (2015). Design of a computer-assisted system to automatically detect cell types using ANA IIF images for the diagnosis of autoimmune diseases. *Journal of Medical Systems, 39*(10), 1–12.

Cheng, C., Chiang, K., & Chen, M. (2016). Intermittent demand forecasting in a tertiary pediatric intensive care unit. *Journal of Medical Systems, 40*(10), 1–12.

Cheng, L., Hu, Y., & Chiou, S. (2017). Applying the temporal abstraction technique to the prediction of chronic kidney disease progression. *Journal of Medical Systems, 41*(5), 1–12.

Chiang, H., & Pao, S. (2016). An EEG-based fuzzy probability model for early diagnosis of Alzheimer's disease. *Journal of Medical Systems, 40*(5), 1–9.

Cho, G., Lee, S., & Lee, T. (2015). An optimized compression algorithm for real-time ECG data transmission in wireless network of medical information systems. *Journal of Medical Systems, 39*(1), 1–8.

Choi, J., Choi, C., Ko, H., & Kim, P. (2016). Intelligent healthcare service using health lifelog analysis. *Journal of Medical Systems, 40*(8), 1–10.

Chou, H., Lin, I.-C., Woung, L., & Tsai, M. (2012). Engagement in E-learning opportunities: An empirical study on patient education using expectation confirmation theory. *Journal of Medical Systems, 36*(3), 1697–1706.

Cicirelli, F., Fortino, G., Giordano, A., Guerrieri, A., Spezzano, G., & Vinci, A. (2016). On the design of smart homes: A framework for activity recognition in home environment. *Journal of Medical Systems, 40*(9), 1–17.

De, L. T., Martínez-Pérez, B., López-Coronado, M., Díaz, J. R., & López, M. M. (2015). Decision support systems and applications in ophthalmology: Literature and commercial review focused on mobile apps. *Journal of Medical Systems, 39*(1), 1–10.

Deng, W., Zhao, H., Zou, L., Li, Y., & Li, Z. (2012). Research on application information system integration platform in medicine manufacturing enterprise. *Journal of Medical Systems, 36*(4), 2289–2295.

Diz, J., Marreiros, G., & Freitas, A. (2016). Applying data mining techniques to improve breast cancer diagnosis. *Journal of Medical Systems, 40*(9), 1–7.

Dobalian, A., Claver, M. L., Pevnick, J. M., Stutsman, H. R., Tomines, A., & Fu, P. (2012). Organizational challenges in developing one of the nationwide health information network trial implementation awardees. *Journal of Medical Systems, 36*(2), 933–940.

Don, S., Chung, D., Choi, E., & Min, D. (2013). An awareness approach to analyze ECG streaming data. *Journal of Medical Systems, 37*(2), 1–9901.

Erinjeri, J., Picus, D., Prior, F., Rubin, D., & Koppel, P. (2009). Development of a Google-based search engine for data mining radiology reports. *Journal for Digital Imaging, 22*, 348–356.

Fallah, M., & Niakan Kalhori, S. R. (2017). A systematic review of data mining applications in patient-centered mobile-based information systems. *Healthcare Informatics Research, 23*(4), 262–270.

Fan, M., Sun, J., Zhou, B., & Chen, M. (2016). The smart health initiative in China: The case of Wuhan, Hubei Province. *Journal of Medical Systems, 40*(3), 1–17.

Faria, B. M., Gonçalves, J., Reis, L. P., & Rocha, Á. (2015). A clinical support system based on quality of life estimation. *Journal of Medical Systems, 39*(10), 1–11.

Faust, O., Rajendra, A. U., Ng, E. Y. K., Ng, K., & Suri, J. S. (2012). Algorithms for the automated detection of diabetic retinopathy using digital fundus images: A review. *Journal of Medical Systems, 36*(1), 145–157.

Feldman, B., Martin, E. M., & Skotnes, T. (2012). Big data in healthcare hype and hope. *Dr. Bonnie, 360*, 122–125.

Figueroa, R. L., & Flores, C. A. (2016). Extracting information from electronic medical records to identify the obesity status of a patient based on comorbidities and bodyweight measures. *Journal of Medical Systems, 40*(8), 1–9.

Forsvik, H., Voipio, V., Lamminen, J., Doupi, P., Hyppönen, H., & Vuokko, R. (2017). Literature review of patient record structures from the physician's perspective. *Journal of Medical Systems, 41*(2), 1–10.

Foster, R. (2014). Health Care Big Data is a big opportunity to address data overload. *Matchcite*. http://www.zdnet.com/blog/health/big-data meets-medical-analysis-video/500. Accessed 28 Sept 2014.

Gambhir, S., Malik, S. K., & Kumar, Y. (2016). Role of soft computing approaches in HealthCare domain: A mini review. *Journal of Medical Systems, 40*(12), 1–20.

Garrison, L. P. (2013). Universal health coverage—Big thinking versus big data. *The Value in Health, 16*(1), S1–S3.

Grams, R. (2012a). American medical informatics review for 2011. *Journal of Medical Systems, 36*(2), 363–366.

Grams, R. (2012b). In the world of medical alphabet soup – "will a workable EMR or EHR please stand up?". *Journal of Medical Systems, 36*(5), 3079–3081.

Grams, R. (2012c). The progress of an American EHR-Part 1. *Journal of Medical Systems, 36*(5), 3077–3078.

Guédon, A. C., Paalvast, M., Meeuwsen, F. C., Tax, D. M., van Dijke, A. P., Wauben, L. S., van der Elst, M., Dankelman, J., & van den Dobbelsteen, J. J. (2016). 'It is time to prepare the next patient' real-time prediction of procedure duration in laparoscopic cholecystectomies. *Journal of Medical Systems, 40*(12), 1–6.

Guo, P., Wang, J., Ji, S., Geng, X. H., & Xiong, N. N. (2015). A lightweight encryption scheme combined with trust management for privacy-preserving in body sensor networks. *Journal of Medical Systems, 39*(12), 1–8.

Guo, C., Zhuang, R., Jie, Y., Ren, Y., Wu, T., & Choo, K. R. (2016). Fine-grained database field search using attribute-based encryption for E-healthcare clouds. *Journal of Medical Systems, 40*(11), 1–8.

Gurupur, V. P., Suh, S. C., Selvaggi, R. R., Karla, P. R., Nair, J. S., & Ajit, S. (2012). An approach for building a personal health information system using conceptual domain knowledge. *Journal of Medical Systems, 36*(6), 3685–3693.

Harpaz, R., Vilar, S., DuMouchel, W., Salmasian, H., Haerian, K., Shah, N., Chase, H., & Friedman, C. (2012). Combing signals from spontaneous reports and electronic health records for detection of adverse drug reactions. *Journal of American Informatics Association, 20*, 413–419.

Hristovski, D., Kastrin, A., Dinevski, D., Burgun, A., Iberna, L., & Rindfleisch, T. C. (2016). Using literature-based discovery to explain adverse drug effects. *Journal of Medical Systems, 40*(8), 1–5.

Hsiao, T., Wu, Z., Chung, Y., Chen, T., & Horng, G. (2012). A secure integrated medical information system. *Journal of Medical Systems, 36*(5), 3103–3113.

Hsieh, S., Cheng, P., Chen, C., Huang, K., Chen, P., Weng, Y., Hsieh, S., & Lai, F. (2010). A multi-voting enhancement for newborn screening healthcare information system. *Journal of Medical Systems, 34*(4), 727–733.

Hsieh, N., Chang, C., Lee, K., Chen, J., & Chan, C. (2012). Technological innovations in the development of cardiovascular clinical information systems. *Journal of Medical Systems, 36*(2), 965–978.

Hsu, W., & Pan, J. (2013). The secure authorization model for healthcare information system. *Journal of Medical Systems, 37*(5), 1–9974.

Huang, B., Zhu, P., & Wu, C. (2012a). Customer-centered careflow modeling based on guidelines. *Journal of Medical Systems, 36*(5), 3307–3319.

Huang, Z., Lu, X., & Duan, H. (2012b). Using recommendation to support adaptive clinical pathways. *Journal of Medical Systems, 36*(3), 1849–1860.

Huang, Z., Lu, X., & Duan, H. (2013). Latent treatment pattern discovery for clinical processes. *Journal of Medical Systems, 37*(2), 1–9915.

Huang, Z., Dong, W., Ji, L., & Duan, H. (2016). Outcome prediction in clinical treatment processes. *Journal of Medical Systems, 40*(1), 1–13.

Islam, M. S., Hasan, M. M., Wang, X., Germack, H. D., & Noor-E-Alam, M. (2018). A systematic review on healthcare analytics: Application and theoretical perspective of data mining. *Healthcare (Basel), 6*(2), 54.

Issac Niwas, S., Palanisamy, P., Chibbar, R., & Zhang, W. J. (2012). An expert support system for breast cancer diagnosis using color wavelet features. *Journal of Medical Systems, 36*(5), 3091–3102.

Karabulut, E. M., & Ibrikci, T. (2014). Effective automated prediction of vertebral column pathologies based on logistic model tree with SMOTE preprocessing. *Journal of Medical Systems, 38*(5), 1–50.

Karimi, S., Wang, C., Metke-Jimenez, A., Gaire, R., & Paris, C. (2015). Text and data mining techniques in adverse drug reaction detection. *ACM Computer Survey, 47*(56), 1–39.

Karla, P. R., & Gurupur, V. P. (2013). C-PHIS: A concept map-based knowledge base framework to develop personal health information systems. *Journal of Medical Systems, 37*(5), 1–9970.

Keltch, B., Lin, Y., & Bayrak, C. (2014). Comparison of AI techniques for prediction of liver fibrosis in hepatitis patients. *Journal of Medical Systems, 38*(8), 1–60.

Keramidas, E. G., Maroulis, D., & Iakovidis, D. K. (2012). tauND: A thyroid nodule detection system for analysis of ultrasound images and videos. *Journal of Medical Systems, 36*(3), 1271–1281.

Khan, W. A., Khattak, A. M., Hussain, M., Amin, M. B., Afzal, M., Nugent, C., & Lee, S. (2014). An adaptive semantic based mediation system for data interoperability among health information systems. *Journal of Medical Systems, 38*(8), 1–28.

Khennak, I., & Drias, H. (2017). Bat-inspired algorithm based query expansion for medical web information retrieval. *Journal of Medical Systems, 41*(2), 1–16.

Kohlmann, M., Gietzelt, M., Jähne-Raden, N., Marschollek, M., Song, B., Wolf, K., & Haux, R. (2014). A collaboration tool based on SNOCAP-HET. *Journal of Medical Systems, 38*(1), 1–9996

Korkmaz, S. A., & Poyraz, M. (2014). A new method based for diagnosis of breast cancer cells from microscopic images: DWEE – JHT. *Journal of Medical Systems, 38*(9), 1–92.

Koutkias, V., & Jaunt, M. (2016). A multiagent system for integrated detection of pharmacovigilance signals. *Journal of Medical Systems, 40*(2), 1–14.

Koyuncugil, A. S., & Ozgulbas, N. (2010a). Detecting road maps for capacity utilization decisions by clustering analysis and CHAID decision tress. *Journal of Medical Systems, 34*(4), 459–469.

Koyuncugil, A. S., & Ozgulbas, N. (2010b). Donor research and matching system based on data mining in organ transplantation. *Journal of Medical Systems, 34*(3), 251–259.

Koyuncugil, A. S., & Ozgulbas, N. (2012). Early warning system for financially distressed hospitals via data mining application. *Journal of Medical Systems, 36*(4), 2271–2287.

Kozat, S. S., Vlachos, M., Lucchese, C., VAN Herle, H., & Yu, P.S. (2009). Embedding and retrieving private metadata in electrocardiograms. *Journal of Medical Systems, 33*(4), 241–259.

Kuo, M. H., Kushniruk, A., & Borycki, E. (2011). A comparison of national health data interoperability approaches in Taiwan, Denmark, and Canada.

Kuo, M. H., Sahama, T., Kushniruk, A. W., Borycki, E. M., & Grunwell, D. K. (2014). Health big data analytics: Current perspectives, challenges, and potential solutions. *International Journal of Big Data Intelligence, 1*(1–2), 114–126.

Kuo, R. J., Huang, M. H., Cheng, W. C., Lin, C. C., & Wu, Y. H. (2015). Application of a two-stage fuzzy neural network to a prostate cancer prognosis system. *Artificial Intelligence in Medicine, 63*(2), 119–133. https://doi.org/10.1016/j.artmed.2014.12.008. Epub 2014 Dec 30. PMID: 25576196.

Kupusinac, A., Stokic, E., & Kovacevic, I. (2016). Hybrid EANN-EA system for the primary estimation of cardiometabolic risk. *Journal of Medical Systems, 40*(6), 1–9.

Lalos, A. S., Lakoumentas, J., Dimas, A., & Moustakas, K. (2016). Energy efficient monitoring of metered dose inhaler usage. *Journal of Medical Systems, 40*(12), 1–10.

Latif, R., Abbas, H., & Assar, S. (2014). Distributed denial of service (DDoS) attack in cloud-assisted wireless body area networks: A systematic literature review. *Journal of Medical Systems, 38*(11), 1–10.

Lavra˘c, N., Bohanec, M., Pur, A., Cestnik, B., Debeljak, M., & Kobler, A. (2007). Data mining and visualization for decision support and modeling of public health-care resources. *Journal of Biomedicine Informatics, 40*, 438–447.

Lee, C., Hsu, C., Lai, Y., & Vasilakos, A. (2013). An enhanced mobile-healthcare emergency system based on extended chaotic maps. *Journal of Medical Systems, 37*(5), 1–9973.

Lee, H., Ahn, H., Choi, S., & Choi, W. (2014). The SAMS: Smartphone addiction management system and verification. *Journal of Medical Systems, 38*(1), 1.

León, M. C., Nieto-Hipólito, J. I., Garibaldi-Beltrán, J., Amaya-Parra, G., Luque-Morales, P., Magaña-Espinoza, P., & Aguilar-Velazco, J. (2016). Designing a model of a digital ecosystem for healthcare and wellness using the business model canvas. *Journal of Medical Systems, 40*(6), 1–9.

Li, J., Zhang, X., Chu, J., Suzuki, M., & Araki, K. (2012). Design and development of EMR supporting medical process management. *Journal of Medical Systems, 36*(3), 1193–1203.

Lin, K., & Hsieh, Y. (2015). Classification of medical datasets using SVMs with hybrid evolutionary algorithms based on endocrine-based particle swarm optimization and artificial bee colony algorithms. *Journal of Medical Systems, 39*(10), 1–9.

Lin, C., Chen, T., Tsai, H., Lee, W., Hsu, T., & Kao, Y. (2015). A novel anti-classification approach for knowledge protection. *Journal of Medical Systems, 39*(10), 1–10.

Lin, C., Song, Z., Song, H., Zhou, Y., Wang, Y., & Wu, G. (2016a). Differential privacy preserving in big data analytics for connected health. *Journal of Medical Systems, 40*(4), 1–9.

Lin, C., Pao, C., Chen, Y., Liu, C., & Hsu, H. (2016b). Ellipsis and coreference resolution in a computerized virtual patient dialogue system. *Journal of Medical Systems, 40*(9), 1–15.

López, M. M., López, M. M., de la Torre Díez, I., Jimeno, J. C., & López-Coronado, M. (2016). A mobile decision support system for red eye diseases diagnosis: Experience with medical students. *Journal of Medical Systems, 40*(6), 1–10.

López, M. M., López, M. M., de la Torre Díez, I., Jimeno, J. C. P., & López-Coronado, M. (2017). mHealth app for iOS to help in diagnostic decision in ophthalmology to primary care physicians. *Journal of Medical Systems, 41*(5), 1–7.

López-Nava, I. H., Arnrich, B., Muñoz-Meléndez, A., & Güneysu, A. (2017). Variability analysis of therapeutic movements using wearable inertial sensors. *Journal of Medical Systems, 41*(1), 1–19.

Luo, G. (2013). Open issues in intelligent personal health record - an updated status report for 2012. *Journal of Medical Systems, 37*(3), 1–9943.

Luo, G. (2014). A roadmap for designing a personalized search tool for individual healthcare providers. *Journal of Medical Systems, 38*(2), 1–6.

Luo, G., & Grams, R. (2012). 1st ACM international health informatics symposium (IHI). *Journal of Medical Systems, 36*(2), 367–370.

Lv, Z., Chirivella, J., & Gagliardo, P. (2016). Bigdata oriented multimedia mobile health applications. *Journal of Medical Systems, 40*(5), 1–10.

Magalhães, T., Lopes, S., Gomes, J., & Seixo, F. (2016). The predictive factors on extended hospital length of stay in patients with AMI: laboratory and administrative data. *Journal of Medical Systems, 40*(1), 1–7.

Matsoukas, P., Williams, R., Davies, C., Ainsworth, J., & Buchan, I. (2015). User interface requirements for web-based integrated care pathways: Evidence from the evaluation of an online care pathway investigation tool. *Journal of Medical Systems, 39*(11), 1–15.

Mehmood, I., Sajjad, M., & Baik, S. W. (2014). Video summarization based tele-endoscopy: A service to efficiently manage visual data generated during wireless capsule endoscopy procedure. *Journal of Medical Systems, 38*(9), 1–109.

Melillo, P., Orrico, A., Scala, P., Crispino, F., & Pecchia, L. (2015). Cloud-based smart health monitoring system for automatic cardiovascular and fall risk assessment in hypertensive patients. *Journal of Medical Systems, 39*(10), 1–7.

Mezghani, E., Exposito, E., Drira, K., DA Silveira, M., & Pruski, C. (2015). A semantic big data platform for integrating heterogeneous wearable data in healthcare. *Journal of Medical Systems, 39*(12), 1–8.

Monteiro, E., Costa, C., & Oliveira, J. L. (2017). A de-identification pipeline for ultrasound medical images in DICOM format. *Journal of Medical Systems, 41*(5), 1–16.

Moola, S., Munn, Z., Tufanaru, C., Aromataris, E., Sears, K., Sfetcu, R., Currie, M., Qureshi, R., Mattis, P., Lisy, K., & Mu, P.-F. (2020). Chapter 7: Systematic reviews of etiology and risk. In E. Aromataris & Z. Munn (Eds.), *JBI manual for evidence synthesis*. JBI. Available from https://synthesismanual.jbi.global

Mudumbai, S., Ayer, F., & Stefanko, J. (2017). Perioperative and ICU healthcare analytics within a veterans integrated system network: A qualitative gap analysis. *Journal of Medical Systems, 41*(8), 1–8.

Mueen, A., Zainuddin, R., & Sapiyan Baba, M. (2010). MIARS: A medical image retrieval system. *Journal of Medical Systems, 34*(5), 859–864.

Neves, J., Martins, M. R., Vilhena, J., Neves, J., Gomes, S., Abelha, A., Machado, J., & Vicente, H. (2015). A soft computing approach to kidney diseases evaluation. *Journal of Medical Systems, 39*(10), 1–9.

Ng, E. Y. K., Rajendra, A. U., & Suri, J. (2011). Topic of special issue: Distributed diagnosis and home healthcare. *Journal of Medical Systems, 35*(5), 825–827.

Nguyen, T., Khosravi, A., Creighton, D., & Nahavandi, S. (2015). Classification of healthcare data using the genetic fuzzy logic system and wavelets. *Expert Systems with Application, 42*, 2184–2197.

Niaksu, O., Skinulyte, J., & Duhaze, H. G. (2014). A systematic literature review of data mining applications in healthcare. In Z. Huang, C. Liu, J. He, & G. Huang (Eds.), *Web information systems engineering – WISE 2013 workshops. WISE 2013* (Lecture notes in computer science) (Vol. 8182). Springer.

Niwas, S. I., Lin, W., Bai, X., Kwoh, C. K., Sng, C. C., Aquino, M. C., Chew, P. T., & K. (2015). Reliable feature selection for automated angle closure glaucoma mechanism detection. *Journal of Medical Systems, 39*(3), 1–10.

Nowaková, J., Prílepok, M., & Snášel, V. (2017). Medical image retrieval using vector quantization and fuzzy S-tree. *Journal of Medical Systems, 41*(2), 1–16.

Oh, T. O., & Teege, G. (2011). Using information technology for improved pharmaceutical care delivery in developing countries. Study case: Benin. *Journal of Medical Systems, 35*(5), 1123–1134.

Oliveira, A., Faria, B. M., Gaio, A. R., & Reis, L. P. (2017). Data mining in HIV-AIDS surveillance system. *Journal of Medical Systems, 41*(4), 1–12.

Osborne, T. F., Clark, R. H., Blackowiak, J., Williamson, P. J., Werb, S. M., & Strong, B. W. (2017). Efficiency analysis of an interoperable healthcare operations platform. *Journal of Medical Systems, 41*(4), 1–7.

Palacio, C., Harrison, J. P., & Garets, D. (2010). Benchmarking electronic medical records initiatives in the US: A conceptual model. *Journal of Medical Systems, 34*(3), 273–279.

Panagiotakopoulos, T., Lyras, D., Livaditis, M., Sgarbas, K., Anastassopoulos, G., & Lymberopoulos, D. (2010). A contextual data mining approach toward assisting the treatment of anxiety disorders. *IEEE Transactions on Information Technology in Biomedicine, 14*, 567–581.

Pérez, J., Iturbide, E., Olivares, V., Hidalgo, M., Martínez, A., & Almanza, N. (2015). A data preparation methodology in data mining applied to mortality population databases. *Journal of Medical Systems, 39*(11), 1–6.

Pollettini, J. T., Panico, S. R. G., Daneluzzi, J. C., Tinós, R., Baranauskas, J. A., & Macedo, A. A. (2012). Using machine learning classifiers to assist healthcare-related decisions: Classification of electronic patient records. *Journal of Medical Systems, 36*(6), 3861–3874.

Praveenkumar, B., Suresh, K., Nikhil, A., Rohan, M., Nikhila, B., Rohit, C., & Srinivas, A. (2014). Geospatial technology in disease mapping, E-surveillance and health care for rural population in South India. *International Archives of the Photogrammetry, Remote Sensing and Spatial Information Sciences, 40*(221), 21–26.

Preve, N. (2011). Ubiquitous healthcare computing with sensor grid enhancement with data management system (SEGEDMA). *Journal of Medical Systems, 35*(6), 1375–1392.

Priyanka, K., & Kulennavar, N. (2014). A survey on big data analytics in healthcare. *International Journal of Computer Science and Information Technologies, 5*(4), 5865–5868.

Pustisek, M. (2017). A system for multi-domain contextualization of personal health data. *Journal of Medical Systems, 41*(1), 1–6.

Rafe, V., & Hajvali, M. (2014). A reliable architectural style for designing pervasive healthcare systems. *Journal of Medical Systems, 38*(9), 1–86.

Rafferty, J., Nugent, C., Liu, J., & Chen, L. (2015). Automatic metadata generation through analysis of narration within instructional videos. *Journal of Medical Systems, 39*(9), 1–7.

Raghupathi, W., & Raghupathi, V. (2014). Big data analytics in healthcare: promise and potential. *Healthy Information Science System, 2*(3). https://doi.org/10.1186/2047-2501-2-3

Ramanan, S. V., Radhakrishna, K., Waghmare, A., Raj, T., Nathan, S. P., Sreerama, S. M., & Sampath, S. (2016). Dense annotation of free-text critical care discharge summaries from an Indian hospital and associated performance of a clinical NLP annotator. *Journal of Medical Systems, 40*(8), 1–9.

Ramos, M. I., Cubillas, J. J., & Feito, F. R. (2016). Improvement of the prediction of drugs demand using spatial data mining tools. *Journal of Medical Systems, 40*(1), 1–9.

Rios-Alvarado, A., Lopez-Arevalo, I., Tello-Leal, E., & Sosa-Sosa, V. (2015). An approach for learning expressive ontologies in medical domain. *Journal of Medical Systems, 39*(8), 1–15.

Rorís, V. M., Gago, J. M., Sabucedo, L. Á., Merino, M. R., & Valero, J. S. (2016). An ICT-based platform to monitor protocols in the healthcare environment. *Journal of Medical Systems, 40*(10), 1–7.

Roy Chowdhury, S., Chakrabarti, D., & Saha, H. (2009). Medical diagnosis using adaptive perceptive particle swarm optimization and its hardware realization using field programmable gate array. *Journal of Medical Systems, 33*(6), 447–465.

Sahin, Y. G., & Celikkan, U. (2012). MEDWISE: An innovative public health information system infrastructure. *Journal of Medical Systems, 36*(3), 1719–1729.

Sareen, S., Sood, S. K., & Gupta, S. K. (2016). An automatic prediction of epileptic seizures using cloud computing and wireless sensor networks. *Journal of Medical Systems, 40*(11), 1–18.

Sayyad Shirabad, J., Wilk, S., Michalowski, W., & Farion, K. (2012). Implementing an integrative multi-agent clinical decision support system with open source software. *Journal of Medical Systems, 36*(1), 123–137.

Shah, N. H., & Tenenbaum, J. D. (2012). FOCUS on translational bioinformatics: The coming age of data-driven medicine: Translational bioinformatics' next frontier. *Journal of the American Medical Informatics Association: JAMIA, 19*(e1), e2.

Shi, X., Li, W., Song, J., Hossain, M. S., Mizanur Rahman, S. M., & Alelaiwi, A. (2016). Towards interactive medical content delivery between simulated body sensor networks and practical data center. *Journal of Medical Systems, 40*(10), 1–11.

Singh, J. (2013). Critical appraisal skills programme. *Journal of Pharmacology and Pharmacotherapeutics, 4*(1), 76.

Suciu, G., Suciu, V., Martian, A., Craciunescu, R., Vulpe, A., Marcu, I., Halunga, S., & Fratu, O. (2015). Big data, internet of things and cloud convergence - an architecture for secure E-health applications. *Journal of Medical Systems, 39*(11), 1–8.

Sufi, F., Khalil, I., & Mahmood, A. (2011). Compressed ECG biometric: A fast, secured and efficient method for identification of CVD patient. *Journal of Medical Systems, 35*(6), 1349–1358.

Thakur, R., Hsu, S. H., & Fontenot, G. (2012). Innovation in healthcare: Issues and future trends. *Journal of Business Research, 65*(4), 562–569.

Ting, S. L., Kwok, S. K., Tsang, A. H., & Lee, W. B. (2011). Critical elements and lessons learnt from the implementation of an RFID-enabled healthcare management system in a medical organization. *Journal of Medical Systems, 35*(4), 657–669.

Topan, A., Bayram, D., Özendi, M., Cam, A., Öztürk, Ö., Ayyildiz, T., Kulakçi, H., & Veren, F. (2016). Determination of spatial distribution of children treated in children oncology clinic with the aid of geographic information systems. *Journal of Medical Systems, 40*(10), 1–8.

Triantafyllopoulos, D., Korvesis, P., Mporas, I., & Megalooikonomou, V. (2016). Real-time management of multimodal streaming data for monitoring of epileptic patients. *Journal of Medical Systems, 40*(3), 1–11.

Tsai, M., & Yu, S. (2016). Distance metric based oversampling method for bioinformatics and performance evaluation. *Journal of Medical Systems, 40*(7), 1–9.

Tsai, M., Wang, H., Lee, G., Lin, Y., & Chiu, S. (2016). A decision tree based classifier to analyze human ovarian cancer cDNA microarray datasets. *Journal of Medical Systems, 40*(1), 1–8.

Übeyli, E. D., & Dodu, E. (2010). Automatic detection of erythemato-squamous diseases using k-means clustering. *Journal of Medical Systems, 34*(2), 179–184.

Vanopstal, K., vander Stichele, R., Laureys, G., & Buysschaert, J. (2011). Vocabularies and retrieval tools in biomedicine: Disentangling the terminological knot. *Journal of Medical Systems, 35*(4), 527–543.

Verma, L., Srivastava, S., & Negi, P. C. (2016). A hybrid data mining model to predict coronary artery disease cases using non-invasive clinical data. *Journal of Medical Systems, 40*(7), 1–7.

Vilhena, J., Rosário Martins, M., Vicente, H., Grañeda, J. M., Caldeira, F., Gusmão, R., Neves, J., & Neves, J. (2017). An integrated soft computing approach to Hughes syndrome risk assessment. *Journal of Medical Systems, 41*(3), 1–12.

Villarreal, V., Hervas, R., Fontecha, J., & Bravo, J. (2015). Mobile monitoring framework to design parameterized and personalized m-health applications according to the patient's diseases. *Journal of Medical Systems, 39*(10), 1–6.

Villarreal, V., Hervás, R., & Bravo, J. (2016). A systematic review for mobile monitoring solutions in M-health. *Journal of Medical Systems, 40*(9), 1–12.

Wan, T. T., & H. (2006). Healthcare informatics research: From data to evidence-based management. *Journal of Medical Systems, 30*(1), 3–7.

Wanderer, J. P., Nelson, S. E., Ehrenfeld, J. M., Monahan, S., & Park, S. (2016). Clinical data visualization: The current state and future needs. *Journal of Medical Systems, 40*(12), 1–9.

Wang, Y., Tian, Y., Tian, L., Qian, Y., & Li, J. (2015). An electronic medical record system with treatment recommendations based on patient similarity. *Journal of Medical Systems, 39*(5), 1–9.

Wang, F., Wang, H., Xu, K., Raymond, R., Chon, J., Fuller, S., & Debruyn, A. (2016). Regional level influenza study with geo-tagged twitter data. *Journal of Medical Systems, 40*(8), 1–8.

Yang, F., Lee, A. J., & Kuo, S. (2016a). Mining health social media with sentiment analysis. *Journal of Medical Systems, 40*(11), 1–8.

Yang, Z., Zhou, Q., Lei, L., Zheng, K., & Xiang, W. (2016b). An IoT-cloud based wearable ECG monitoring system for smart healthcare. *Journal of Medical Systems, 40*(12), 1–11.

Yao, Q., Tian, Y., Li, P., Tian, L., Qian, Y., & Li, J. (2015). Design and development of a medical big data processing system based on Hadoop. *Journal of Medical Systems, 39*(3), 1–11.

Yeh, W.-C., Chang, W.-W., & Chung, Y. (2009). A new hybrid approach for mining breast cancer pattern using discrete particle swarm optimization and statistical method. *Expert System, 36*, 8204–8211.

Youngstrom, N. (2012). Programs to reduce readmissions could increase the risk of fraud and abuse, 21 REPORT ON MEDICARE COMPLIANCE 1.

Youssef, A. E. (2014). A framework for secure healthcare systems based on big data analytics in mobile cloud computing environments. *International Journal of Ambient Systems and Applications, 2*(2), 1–11.

Yu, D., Blocker, R. C., Sir, M. Y., Hallbeck, M. S., Hellmich, T. R., Cohen, T., Nestler, D. M., & Pasupathy, K. S. (2016). Intelligent emergency department: Validation of sociometers to study workload. *Journal of Medical Systems, 40*(3), 1–12.

Yue, X., Wang, H., Jin, D., Li, M., & Jiang, W. (2016). Healthcare data gateways: Found healthcare intelligence on blockchain with novel privacy risk control. *Journal of Medical Systems, 40*(10), 1–8.

Zhang, Y., Sun, Y., Phillips, P., Liu, G., Zhou, X., & Wang, S. (2016). A multilayer perceptron based smart pathological brain detection system by fractional Fourier entropy. *Journal of Medical Systems, 40*(7), 1–11.

Zhou, X., Chen, S., Liu, B., Zhang, R., Wang, Y., Li, P., Guo, Y., Zhang, H., Gao, Z., & Yan, X. (2010). Development of traditional Chinese medicine clinical data warehouse for medical knowledge discovery and decision support. *Artificial Intelligence Medicine, 48*, 139–152.

Chapter 7
Malaria Detection Using Machine Learning

Aml Kamal Osman Babikir and Christopher Thron

7.1 Introduction

Malaria is a life-threatening disease caused by parasites that are transmitted to people through the bites of infected female Anopheles mosquitoes. It is considered a serious health problem in the tropics (especially in sub-Saharan Africa), with the estimated 228 million cases and 405,000 deaths attributable to malaria worldwide in 2018 (Abdalla et al., 2007).

According to Abdalla et al., the malaria-producing parasites are protozoa of the genus *Plasmodium*, including two common and widespread species (*Plasmodium falciparum* and *Plasmodium vivax*) and two uncommon species (*Plasmodium malaria* and *Plasmodium oval*). *Plasmodium falciparum* is the most widely encountered species in sub-Saharan Africa and is responsible for about 99.7% of the infected cases and 93% of all malaria deaths in Africa. They also state that in Sudan, in particular, malaria is one of the main causes of mortality and the leading cause of child mortality. It is estimated that it has 7.5–10 million cases each year with 35–40 thousand deaths. As of 2021, the most recent breakout of malaria in Sudan was 2019.

Early and accurate diagnosis followed by effective treatment is crucial in the control of malaria (Mbanefo & Kumar, 2020). Currently, there are several different procedures used for malaria diagnosis. Clinical diagnosis based on symptoms (fever,

A. K. O. Babikir (✉)
Department of Computer Science, University of Khartoum, Khartoum, Sudan

C. Thron
Department of Science and Mathematics, Texas A& M University-Central Texas, Killeen, TX, USA
e-mail: thron@tamuct.edu

M. Alloghani et al. (eds.), *Artificial Intelligence for Data Science in Theory and Practice*, Studies in Computational Intelligence 1006,
https://doi.org/10.1007/978-3-030-92245-0_7

chills, muscle aches, tiredness) is unreliable, because the flu and other viruses cause similar symptoms.

The World Health Organization (WHO) recommends that all patients with clinically suspected malaria should have their diagnosis confirmed by further testing before treatment is initiated. The fastest and cheapest of these tests is rapid diagnostic testing (RDT), which requires only a blood sample applied to a test card. Unfortunately, accuracy is still low, with a large proportion of malaria cases going undetected (Fagbamigbe, 2019). In contrast, the "gold standard" for laboratory testing is microscopy, which uses stained blood samples on prepared slides (Cheesbrough, 2005). However, microscopy requires highly trained technicians to prepare the samples and conduct the analysis, as well as high-quality instrumentation and reagents, which may not be available in the lower level facilities of Sudan. This is a significant barrier to overcome, as health infrastructure in Sudan is extremely fractured and under-resourced (Malik et al., 2004). A final alternative is the molecular diagnosis via polymerase chain reaction (PCR) assay, which detects protozoal DNA in the blood. PCR is more sensitive than microscopy but requires even more expensive equipment and materials (Makanjuola & Taylor-Robinson, 2020). The shortage of medical expertise in labs, along with the lack of facilities needed to establish the proper diagnosis in a short time, leads to increments in the rate of diagnostic errors. This represents a major public health problem, considering that wrong or delayed diagnosis will cause more serious harm to patients than any other type of medical error (Balogh et al., 2015; Tehrani et al., 2013).

The remainder of this chapter is organized as follows. Section 7.2 gives a short review of related studies, Sect. 7.3 describes the methods used in the investigation, Sect. 7.4 presents and interprets results, and Sect. 7.5 gives a summary of conclusions and indicates directions for further investigation.

7.2 Review of Related Literature

Several previous applications of deep learning to malaria diagnosis are described in the following.

Kumar et al. (2021) compared the use of three standard CNN architectures (LeNet, AlexNet, and ResNet) to detect malaria parasites in segmented blood cells. After some modifications, the architectures achieved accuracies of 95.50%, 95%, and 92%, respectively.

Kumari et al. (2020) used various techniques including gray level co-occurrence matrix (GCLM), local binary pattern (LBP), and histogram of oriented gradients (HOG) to extract multiple features to differentiate between healthy and infected cells. By applying a *Support Vector Machine* (SVM) classifier, they reached a classification accuracy of 97.93%.

Das et al. (2013) used marker-controlled watershed transformation to segment erythrocytes and to extract features that describe texture and shape size, and the Bayesian learning and support vector machine (SVM) were used to classify the different stages of malaria (*Plasmodium vivax* and *Plasmodium falciparum*). The accuracies of the Bayesian and SVM approaches were 84% and 83.5%, respectively.

Linder et al. (2014) used computer vision applied to digitally scanned Giemsa-stained thin blood films to identify candidate regions for parasites, based on color and object size. From these regions, image features (local binary patterns, local contrast, and scale-invariant feature transform descriptors) were extracted and used as inputs to an SVM classifier that had been previously trained on digital slides from ten patients and validated on six samples. The SVM classifier achieved a sensitivity of 85% and specificity of 99.9%, compared to an average of 92.5% sensitivity and 100% specificity for two trained technicians who examined the same regions.

Park et al. (2016) used quantitative phase images of unstained cells to detect malaria parasite *Plasmodium falciparum* at the trophozoite or subsequent schizont stages. Optical phase thresholds are used to automatically segment the images, and segmented images were refocused and used to extract 23 features, to which three different machine learning techniques were applied: linear discriminant classification (LDC), logistic regression (LR), and k-nearest neighbor (KNN). All three techniques reached accuracy above 99% for detection of schizont stage or late trophozoites, while detection of early trophozoites reached up to 98%.

7.3 Methodology

7.3.1 Datasets Used

Two datasets were used in this implementation. The first dataset was used for training and evaluating the models, while the second dataset consisted of local samples and was used as examples of actual field data. The two datasets are described in more detail in the following.

The dataset used for training and evaluating the models was downloaded from the National Library of Medicine (NLM) website (Jaeger, 2021). The dataset is documented in Rajaraman et al. (2018). Images were obtained from Giemsa-stained thin blood smear slides from 150 *P. falciparum*-infected patients and 50 healthy patients were photographed with the aid of a mobile application that runs on a standard Android smartphone attached to a conventional light microscope. The slides were collected and photographed at Chittagong Medical College Hospital, Bangladesh. A level-set-based algorithm was applied to detect and segment red blood cells, which were then manually annotated by an expert slide reader at the Mahidol-Oxford Tropical Medicine Research Unit in Bangkok, Thailand. The dataset contains a total of 27,558 cell images with equal instances of parasitized and uninfected cells. Figure 7.1 shows a representative sample of cell images from the dataset.

The dataset of local samples utilized images taken from slides prepared and stained using a RAL staining kit (https://www.ral-diagnostics.fr/), at the El-Romi

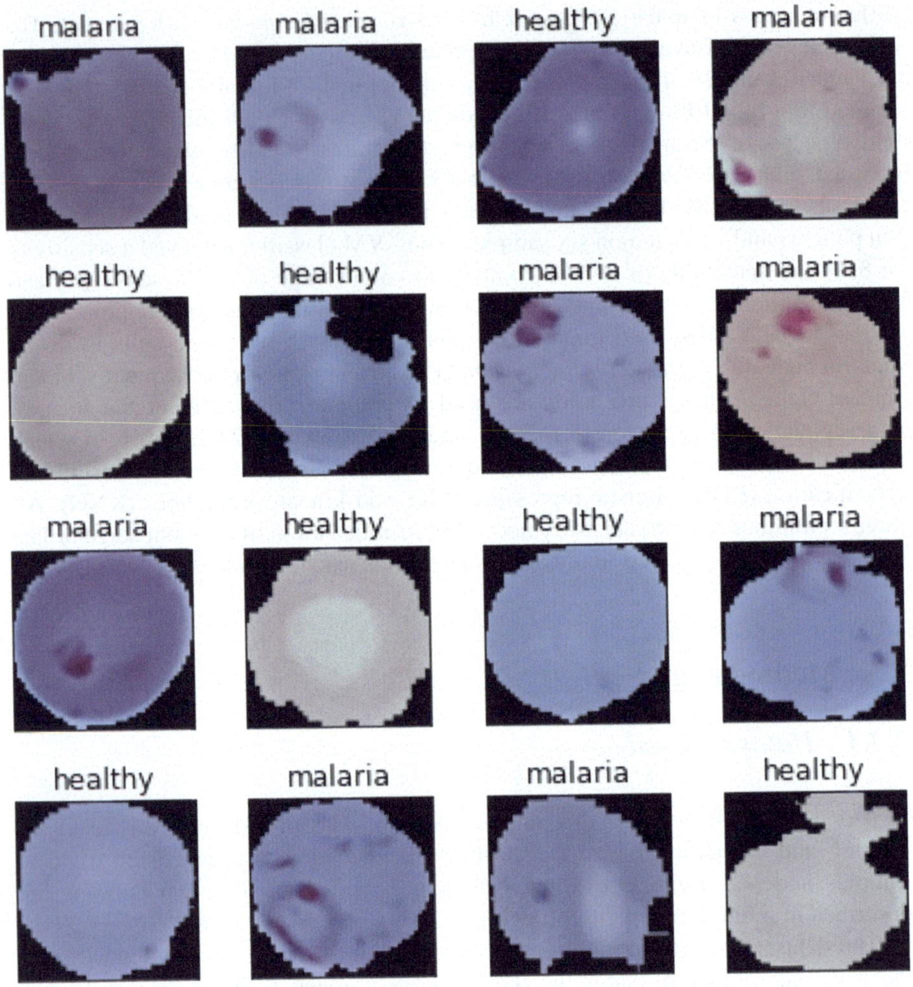

Fig. 7.1 Sample of healthy and infected cells from the NLM database (Jaeger, 2021)

lab in Omdurman, Sudan. Slides were photographed using a digital camera mounted on a microscope and attached to a computer. Images were segmented to cell level using the segmentation method described in the next section. The segmented cells were manually classified by an expert lab technician into infected (positive) cells and healthy (negative) cells. The dataset contains a total of 79 cell images, with 39 infected and 40 healthy cells that were used to train and validate the transfer learning model. Figure 7.2 shows a representative sample of cell images from the dataset.

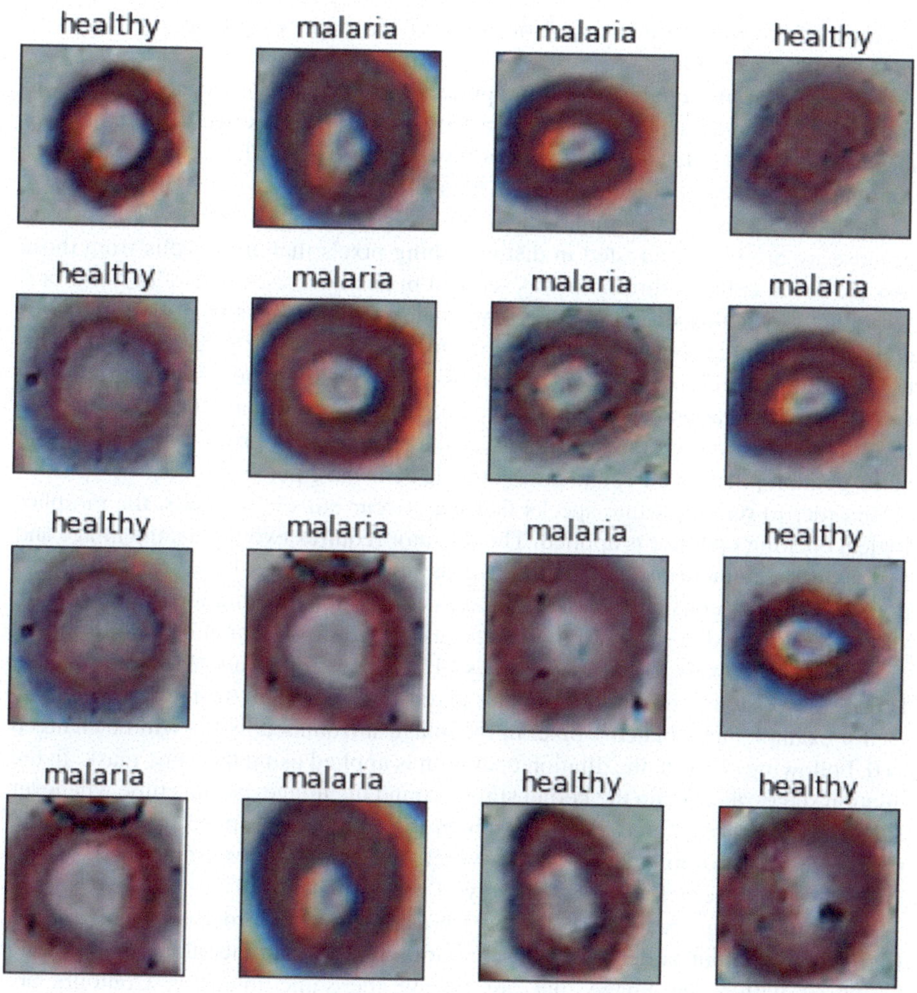

Fig. 7.2 Sample of images of healthy and infected cells from local El Romi lab database

7.3.2 Cell Segmentation Algorithm

The identification procedure first performs image segmentation to isolate cells and then uses machine learning on the segmented cell images to classify cells. In this section, we describe the segmentation procedure, while in the next section we present the machine learning algorithm.

The steps in the segmentation algorithm are as follows:

First, the image is converted to grayscale. Color shades provide little useful information, and grayscale images have less information per pixel. The `cvtColor` command from the Open Source Computer Vision (OpenCV) Python library was used for the conversion.

Next, histogram equalization was applied to the grayscale images. Histogram equalization increases the global contrast of images by effectively spreading out the most frequent intensity values (Dorothy et al., 2015). The `equalizeHist` command was used to obtain histogram equalized images.

The next step is thresholding, which converts pixel values to binary. This is used because we are only interested in distinguishing pixels that are in cells from those that are not in cells. A threshold was set, and pixels were assigned values of 1 or 0 according to whether or not the grayscale value (which ranges from 0 to 255) was above the threshold. For this purpose, the OpenCV function `cv.threshold` was used. The threshold used was calculated using the Otsu method, which minimizes the within-class variance and maximizes between-class variance (Otsu, 1979). The resulting image was also inverted, so that the original dark areas (which represent cells) contain pixels with 1, and areas outside cells have pixels with 0.

In order to remove white specks that may occur outside the cells, the morphological opening operator is applied. The operator requires two inputs: the image and a structuring element or kernel. Both inputs are specified as binary matrices. The opening operator consists of two stages, namely erosion and dilation. In the erosion stage, the kernel slides around the image (as in two-dimensional convolution) and changes image pixels to 0 if the kernel's "1" entries are not all "1"s in the image. Erosion serves to retain 1 values only for pixels that are within the interior of cells—so, for example, an isolated 1 pixel in the image surrounded by 0's will be changed to 0. Following erosion, the dilation operation is applied using the same mask. In the dilation stage, once again the kernel slides around the image, but this time whenever a "1" pixel is encountered all "1" pixels in the mask are changed to "1" pixels in the image. In our implementation, the `MORPH_OPEN` function from OpenCV was applied, and the kernel was a 5×5 array of pixels all set to "1."

After the image is opened, contours are identified in order to localize cells. Formally, a contour is defined as the line joining all the points along a connected boundary within the image that are having the same intensity. Contours are commonly used in shape and size analysis, as well as object detection. To extract contours, OpenCV's `findContours()` function was used, which outputs the contours as a Python list of two-column `NumPy` arrays, where the columns of each array contain the x and y coordinates, respectively, of all points on a single boundary. In our implementation, the `CHAIN_APPROX_NONE` option was used so that all boundary points were stored without compression.

At this point, the different contours have been separated. However, if cells overlap, it is possible for a single contour to enclose multiple cells. These cells can be separated using a variant of the *watershed algorithm* (Beucher & Meyer, 2018). First, individual cells are identified by computing the Euclidean distance from every "1" pixel to the nearest zero pixel and finding peaks (i.e., local maxima) in this distance function. These peaks represent the centers of different cells. To ensure that different peaks represent different cells, peaks that are in adjacent cells are grouped together and assigned the same label using the `ndimage.label` function from the `scipy` library. The peak locations and labels are then passed to the `morphology.watershed` function from the `skimage` library, together with the entire binary image. The watershed algorithm will assign to each "1" pixel a cell label by using the peaks' labels as "seeds" from which to grow individual cells.

At this stage, the pixels have been segmented into separate cells, and all that remains is to crop the segmented cells. This is done by looping over cell labels and finding the smallest bounding rectangle using the `boundingRect` function from OpenCV. Each cropped image is then saved to a separate image file.

7.3.3 Convolutional Neural Network Structure

Convolutional neural networks (CNNs) are commonly used to process 2D pixel data and are used in many powerful image recognition algorithms.

CNNs typically consist of multiple layers, where each layer performs a specific mathematical operation. In the usual case, a series of convolution and pooling operations are performed, followed by a number of fully connected layers with nonlinear activation functions. These different layers are described in more detail below.

Convolution Convolution is a mathematical operation on one- or higher dimensional data that produces local averages so as to extract local features. For image data, the convolution operation involves taking local weighted averages at systematically selected pixel locations in the image. Each weighted average is computed by taking the element-wise multiplication of a matrix consisting of neighboring pixel values with a matrix that represents the convolution kernel and summing the results.

Besides the kernel, two parameters that determine the convolution output are *stride* and *padding*. Stride determines the spacing between pixel locations chosen for local averaging: for example, a stride of three indicates that every third pixel in every third row of pixels is selected for filtering. In other words, a single convolution value is computed for every 3×3 square of pixels in the original image. Padding determines whether or not a border of zeros is placed around the image before filtering. If the border is included, then pixels on the edge of the original image can be filtered; otherwise, filtered pixels must be at least a kernel width's distance from the edge of the image.

In two-dimensional CNN layers, the input data (which often represents images) often has a third dimension (or "depth") which represents different aspects of the image. For example, color images will have a depth of three, corresponding to three color channels (e.g., RGB or HSV). In this case, the kernel will also have the same depth, so, for example, a 3×3 kernel for a color image will actually be a $3 \times 3 \times 3$ array (also called a *tensor*).

In addition to the input depth, multiple kernels may be applied to the same input, so that the layer's output also has a depth. The kernel entries are the key to the layer's function—these entries are adjusted during the training process, so that the convolutional layer picks out features of interest in the input.

As a final step in CNN convolution layers, a bias is added to the convolved values, and the results are then passed through a nonlinear function, (called the *activation function*). Common activation functions include the *Lectified Linear Unit* (ReLU), sigmoid, and hyperbolic tangent functions. Each kernel has its own bias, and biases (like the kernel entries) are adjusted during the training process.

In the `keras` neural network library (which is built on the `tensorflow` platform in Python), convolution is implemented using the `Conv2D` method. Our CNN includes three successive convolutional layers with 3 × 3 kernels, having 32, 64, and 128 output feature maps, respectively, as shown in (Fig. 7.3).

Pooling In CNNs, convolution layers are typically followed by pooling layers. Pooling layers divide each feature map into small rectangles and produce one output per rectangle which represents the important information contained in

Layer (type)	Output Shape	Param #
input_1 (InputLayer)	[(None, 125, 125, 3)]	0
conv2d (Conv2D)	(None, 125, 125, 32)	896
max_pooling2d (MaxPooling2D)	(None, 62, 62, 32)	0
conv2d_1 (Conv2D)	(None, 62, 62, 64)	18496
max_pooling2d_1 (MaxPooling2	(None, 31, 31, 64)	0
conv2d_2 (Conv2D)	(None, 31, 31, 128)	73856
max_pooling2d_2 (MaxPooling2	(None, 15, 15, 128)	0
flatten (Flatten)	(None, 28800)	0
dense (Dense)	(None, 512)	14746112
dropout (Dropout)	(None, 512)	0
dense_1 (Dense)	(None, 512)	262656
dropout_1 (Dropout)	(None, 512)	0
dense_2 (Dense)	(None, 1)	513

```
Total params: 15,102,529
Trainable params: 15,102,529
Non-trainable params: 0
```

Fig. 7.3 Diagram of CNN model generated by keras

that rectangle. This operation reduces the number of parameters (weights and biases), in subsequent layers, thus training time and reducing overfitting. The two most common pooling methods are max pooling and average pooling: CNN uses max pooling, which takes the maximum and average values in each rectangle, respectively. In CNN, max pooling layers with rectangle size 2 × 2 are inserted after each convolution layer, as shown in Fig. 7.3. In `keras`, this is accomplished using the `Maxpooling2D` function.

The output of the final pooling layer is rearranged into a one-dimensional array (this is called "flattening"), and every output is connected to every neuron in the next layer (such a layer is designated as *fully connected*). Our CNN contains a succession of two fully connected layers with 512 neurons per layer, followed by a single-neuron output layer as shown in Fig. 7.3. The final layer uses a sigmoid activation function in order to produce a value between 0 and 1. The sigmoid function $\sigma(x)$ is given by

$$\sigma(x) = \frac{1}{1 + e^x}. \tag{7.1}$$

This output value gives the CNNs estimate of the status of the cell: output values near 1 or 0, respectively, indicate that the CNN has a high confidence that the cell is infected or non-infected, respectively. An output value near 0.5 indicates that the CNN is uncertain of the infective status of the cell.

The overall structure of the CNN shown in Fig. 7.3 can be divided into two parts: the feature extraction part and the classification part. The convolution and pooling layers perform feature extraction, while the subsequent fully connected layers perform classification. (Note that the "dropout" layers shown in the Fig. 7.3 are only used in training and will be explained below.)

7.3.4 *Training of Base Model*

The training process adjusts the weights and biases of the three convolutional layers and three fully connected layers. Training proceeds in multiple epochs, where each epoch has training and validation phases. The parameter adjustment is done during the training phase, and the adjusted parameters are evaluated during the validation phase. Training requires the definition of a loss function, which is used to measure the difference between the CNN's classification into infected/non-infected cells and the true (expert-determined) classification. In CNN, the cross-entropy is used as the loss function, where cross-entropy is defined as

$$\text{Cross entropy} = \frac{1}{N}\sum_{n=1}^{N}(\text{cross entropy for image } n), \tag{7.2}$$

where

$$\text{cross entropy for image } n = \begin{cases} -\log(y_n) & \text{if } \hat{y}_n = 1; \\ -\log(1 - y_n) & \text{if } \hat{y}_n = 0, \end{cases} \quad (7.3)$$

where y_n is the output for the nth training image, and $\hat{y}_n$ is equal to 1 or 0 according to whether the true classification of the nth image is infected or non-infected, respectively.

During the training phase of each epoch, the parameters (weights and biases) of the CNN are incrementally adjusted by running through the testing images in random order, evaluating the loss function for each image and its gradient with respect to the parameters and iteratively nudging the parameters along the negative gradient direction to reduce the loss function. The gradients for the parameters for the different layers are computed using backpropagation Rumelhart et al. (1986). Each "nudge" is computed as the gradient times the *learning rate*, which is an algorithm parameter that typically is gradually reduced during the learning process. In algorithm, the size of the learning rate is controlled automatically using the popular *Adam algorithm* (Kingma & Ba, 2014), which is included as an optimizer option in `keras`.

In Fig. 7.3, each fully connected layer is followed by a "dropout" layer. These dropout layers are inserted only during training. Their purpose is to prevent overfitting. During the calculation of the loss function described in the previous paragraph, each connection between fully connected layers is randomly "dropped" with probability p: this means all the inputs and outputs to this neuron will be disabled at the current iteration. The dropped-out neurons are resampled at every training step, so the dropped-out neurons vary from step to step. The hyperparameter p is called the *dropout rate*.

Following the adjustment of CNN parameters during the training phase, the loss function is computed for a separate set of images in order to validate the CNN's new parameters. For this purpose, the LHNCBC dataset was split into training and validation sets of 19,437 and 8331 images, respectively. This division follows the usual practice of reserving 30% of the images for validation. The entire training process utilized six epochs, where each epoch consisted of parameter optimization (via backpropagation) followed by validation.

7.3.5 Transfer Learning for Training of Specialized Model

The training process described in the previous section tunes the CNN so that it can accurately classify cell images that resemble those in the LHNCBC dataset. However, this study aims to apply the CNN to images from the El-Romi lab in Omdurman, Sudan, which obtained cell images using different procedures. Hence the images of interest have slightly different characteristics from the images used for training the CNN. To address this problem, we apply transfer learning, which is a machine learning technique wherein a model developed for one task is reused as the starting point for a model intended for a different (but related) task. In this way, it is possible to construct a CNN model that both benefits from the extensive training of the original CNN model and is particularly adapted to give good classification results for images from local samples.

To accomplish transfer learning, five more layers are appended to the 13 layers shown in Fig. 7.3. The additional layers are duplicates of the final five layers in the original CNN. This augmented CNN is then retrained on different data (to be described below) over 10 epochs.

The dataset for retraining was based on images taken by the local El-Romi laboratory. However, the number of available images was quite small (a total of 79 images, with 40 healthy and 39 infected) and insufficient for effective retraining. To enlarge the dataset, new images were generated by applying image transformation operations such as rotation, shearing, translation, and zooming. This technique is called *image augmentation* and can be implemented using the `ImageDataGenerator` utility in `keras`.

During the retraining process, the parameters of the first 13 layers are frozen, and only the parameters in the five new layers are allowed to vary. The model was trained over 100 epochs, with accuracy and loss recorded after each epoch. The 79 original images were divided into training and validation sets of size 55 and 24, respectively. Both training and validation sets were augmented as described in the previous paragraph, such that each original image was used to create 39 additional images via image transformations.

7.3.6 Model Summary

The overall system for malaria detection in Sudan makes use of the segmentation algorithms and the CNN described above. In practice, the system operates in three phases: segmentation, identification, and parasitemia level calculation. Images from smears are segmented using the algorithm described in Sect. 7.3. Representative segmented cells are passed to the classification model, where a decision is made on each cell as to whether it is infected by a parasite. The parasitemia level is estimated as the proportion of infected cells in the sample.

7.4 Results

In this section, we describe the accuracy performance of the base and retrained model in predicting their respective datasets.

7.4.1 Base Model Training and Accuracy

Figure 7.4 shows the epoch-by-epoch accuracy and loss for the base CNN during the training process. After 4 epochs, the validation accuracy and loss stop improving, while the training accuracy and loss continue to improve. In order to avoid overfitting, training was stopped after 5 epochs. After training, when applied to the validation set, the positive recall (the proportion of malarial cells that are correctly identified) was 0.94, while the negative recall (the proportion of uninfected cells correctly identified) is 0.96, yielding an overall accuracy of 0.95. It is clear that the base CNN performs well on the given sample.

Figure 7.5 shows the epoch-by-epoch accuracy and loss for the retraining process. We find no consistent epoch-by-epoch improvement. It is clear that the retraining process is not producing any improvements at all in the accuracy, even after 100 epochs. The model is apparently not able to identify any features in the El-Romi images that distinguish malarial from non-malarial cells, despite having

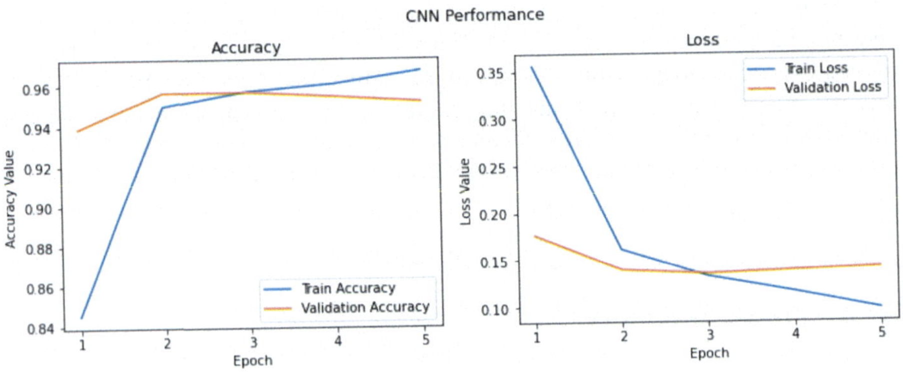

Fig. 7.4 Accuracy and loss for training process for base CNN

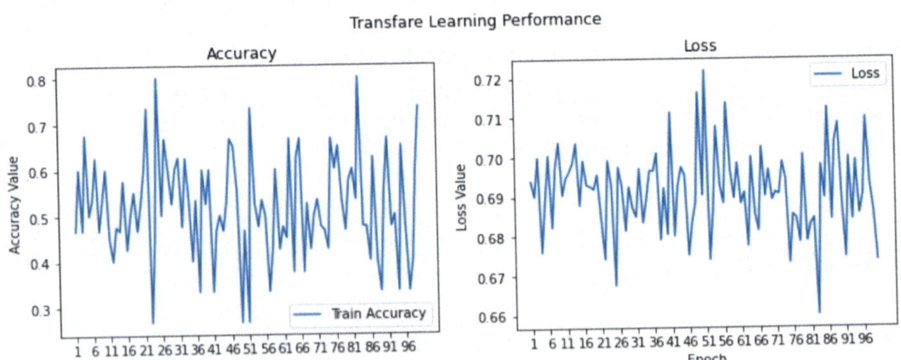

Fig. 7.5 Accuracy and loss for transfer learning for full CNN

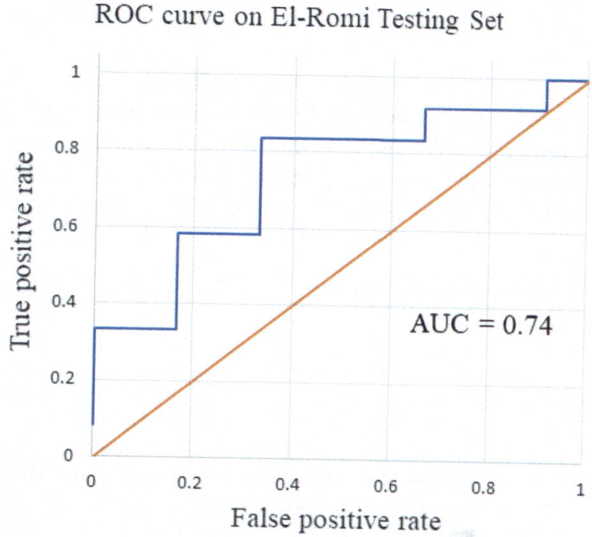

Fig. 7.6 ROC curve for transfer learning model on El-Romi test data

very strong discriminative ability on the NLM database. In fact, on the testing set of 12 infected and 12 non-infected cells, all except one were identified as non-infected, and only one malarial cell was correctly identified. An examination of model outputs from the 24 testing images showed that on a scale from 0 (definitely uninfected) to 1 (definitely infected) with 0.5 indicating the highest uncertainty, all outputs ranged between 0.477 and 0.505 showing that all determinations were made with extremely low confidence. Figure 7.6 shows the receiver operating characteristic (ROC) curve for the model applied to the testing data. As indicated by the figure, the area under the curve (AUC) statistic is 0.74, which seems to indicate good predictive performance. However, this indication is highly misleading, because it presumes that the user can set the choice threshold after looking at the data, which is a kind of cheating. The actual threshold used was 0.5, and all but one of the data points fell below this value.

7.5 Conclusions

Despite the good performance of the base model, it is clear that the retrained model fails to provide useful information in malaria diagnosis. A priori, there are two factors which might plausibly explain this poor performance. First, the quality of the images is poor, and second, the size of the database for retaining is small. In this case, we should put most of the blame on the poor image quality. It is possible that the small retraining training set may have had a minor effect on the accuracy because the training set had one more uninfected than infected cell, meaning that uninfected predictions have a slight probabilistic advantage over infected. However, in general, a small retraining set is liable to cause overfitting, but not poor performance on the training set itself. Furthermore, a visual inspection of Fig. 7.2 shows that the El-Romi cells with and without malaria are extremely difficult to tell apart, even by humans who possess a highly developed visual cortex for processing image information. We conclude that further research in this area should focus on image enhancement techniques, which can lead to better discrimination between malarial and uninfected cells.

References

Abdalla, S. I., Malik, E. M., & Ali, K. M. (2007). The burden of malaria in Sudan: Incidence, mortality and disability–adjusted life–years. *Malaria Journal, 6*(1), 1–9.

Balogh, E. P., Miller, B. T., & Ball, J. R. (2015). Improving diagnosis in health care.

Beucher, S., & Meyer, F. (2018). The morphological approach to segmentation: The watershed transformation. In *Mathematical morphology in image processing* (pp. 433–481). CRC Press.

Cheesbrough, M. (2005). *District laboratory practice in tropical countries* (2nd ed.). Cambridge University Press.

Das, D. K., Ghosh, M., Pal, M., Maiti, A. K., & Chakraborty, C. (2013). Machine learning approach for automated screening of malaria parasite using light microscopic images. *Micron, 45*, 97–106.

Dorothy, R., Joany, R. M., Joseph Rathish, R., Santhana Prabha, S., Rajendran, S., & Joseph, S. (2015). Image enhancement by histogram equalization. *International Journal of Nano Corrosion Science and Engineering, 2*(4), 21–30.

Fagbamigbe, A. F. (2019). On the discriminatory and predictive accuracy of the RDT against the microscopy in the diagnosis of malaria among under-five children in Nigeria. *Malaria Journal, 18*(1), 1–12.

Jaeger, S. (2021). Malaria datasets. Accessed August 25, 2021.

Kingma, D. P., & Ba, J. (2014). Adam: A method for stochastic optimization. Preprint, arXiv:1412.6980.

Kumar, R., Gupta, A., & Mishra, A. (2021). Design of ensemble learning model to diagnose malaria disease using convolutional neural network. In *International Conference on Innovative Computing and Communications* (pp. 1165–1176). Springer.

Kumari, U., Memon, M. M., Narejo, S., & Afzal, M. (2020). Malaria disease detection using machine learning. In *2nd International Conference on Computational Sciences and Technologies (INCCST 20)*.

Linder, N., Turkki, R., Walliander, M., Mårtensson, A., Diwan, V., Rahtu, E., Pietikäinen, M., Lundin, M., & Lundin, J. (2014). A malaria diagnostic tool based on computer vision screening and visualization of *Plasmodium falciparum* candidate areas in digitized blood smears. *PLoS One, 9*(8), e104855.

Makanjuola, R. O., & Taylor-Robinson, A. W. (2020). Improving accuracy of malaria diagnosis in underserved rural and remote endemic areas of sub-Saharan Africa: A call to develop multiplexing rapid diagnostic tests. *Scientifica, 2020*, Article ID 3901409.

Malik, E., Atta, H. Y., Weis, M., Lang, A., Puta, C., Lettenmaier, C., & Bell, A. (2004). Sudan roll back malaria consultative mission: Essential actions to support the attainment of the Abuja targets.

Mbanefo, A., & Kumar, N. (2020). Evaluation of malaria diagnostic methods as a key for successful control and elimination programs. *Tropical Medicine and Infectious Disease, 5*(2), 102.

Otsu, N. (1979). A threshold selection method from gray-level histograms. *IEEE Transactions on Systems, Man, and Cybernetics, 9*(1), 62–66.

Park, H. S., Rinehart, M. T., Walzer, K. A., Chi, J.-T. A., & Wax, A. (2016). Automated detection of p. falciparum using machine learning algorithms with quantitative phase images of unstained cells. *PloS One, 11*(9), e0163045.

Rajaraman, S., Antani, S. K., Poostchi, M., Silamut, K., Hossain, Md. A., Maude, R. J., Jaeger, S., & Thoma, G. R. (2018). Pre-trained convolutional neural networks as feature extractors toward improved malaria parasite detection in thin blood smear images. *PeerJ, 6*, e4568.

Rumelhart, D. E., Hinton, G. E., & Williams, R. J. (1986). Learning representations by back-propagating errors. *Nature, 323*(6088), 533–536

Tehrani, A. S. S., Lee, H. W., Mathews, S. C., Shore, A., Makary, M. A., Pronovost, P. J., & Newman-Toker, D. E. (2013). 25-Year summary of us malpractice claims for diagnostic errors 1986–2010: An analysis from the national practitioner data bank. *BMJ Quality & Safety, 22*(8), 672–680.

Chapter 8
Automatic Number Plate Recognition System for Oman

Mohammed Al Awaimri, Sallam Fageeri, Aiman Moyaid, Christopher Thron, and Abdullah ALhasanat

8.1 Introduction

Due to the rapid worldwide growth of the transportation sector, new technologies are needed to deal with the increasing number of vehicles and increasing complexity of traffic. There is increasing need for electronic systems for traffic management with fast response times and greater adaptability. One key component of such systems is the automatic identification and recognition of vehicles number plates, which is necessary to manage and control vehicle activity (Mubarak et al., 2017). Vehicles in countries worldwide are identified by their number plates. These number plates are visually readable by humans, but not computers. A computerized vehicle plate recognition system requires conversion of visual (image) data to digital data, on which further processing is performed. Automatic number plate recognition (ANPR) is defined as a technology that use character recognition on digitized images to read vehicles registration plate to identify the data of vehicles. It has wide variety of applications in traffic management, including automatic parking of

M. Al Awaimri (✉) · S. Fageeri · A. Moyaid
Department of Information Systems, College of EMIS, University of Nizwa, Nizwa, Sultanate of Oman
e-mail: 21482026@uofn.edu.om; sallam@unizwa.edu.om; aiman.moyaid@unizwa.edu.om

C. Thron
Department of Science and Mathematics, Texas A&M University-Central Texas, Killeen, TX, USA
e-mail: thron@tamuct.edu

A. ALhasanat
Department of Electrical and Computer Engineering, College of EA, University of Nizwa, Nizwa, Sultanate of Oman
e-mail: a.ismail@unizwa.edu.om

M. Alloghani et al. (eds.), *Artificial Intelligence for Data Science in Theory and Practice*, Studies in Computational Intelligence 1006,
https://doi.org/10.1007/978-3-030-92245-0_8

Fig. 8.1 Common types of vehicle number plates in Oman

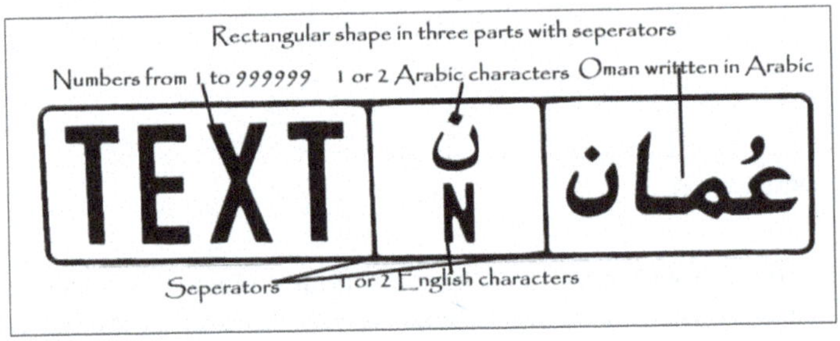

Fig. 8.2 Basic layout of Oman number plates

vehicles, preventing unauthorized vehicle entry, gathering traffic statistics, locating stolen vehicles, performing background checks, control of border crossings and airport entry, traffic monitoring, radar-based speed, and driver seatbelt control (Abbas & Rashid, 2017; Saghaei, 2016; Ghasempour, 2015; Singh, 2016). Despite the great variety of uses, all these different applications of automatic vehicle number plate recognition systems utilize the same basic processing algorithms. Machine learning and deep learning as well as Internet of things (IoT) technology all play important roles in such systems (Saghaei, 2016).

Detecting and extracting characters from digital images poses significant challenges. Poor visibility conditions such as darkness, shadows, rain, or fog can impair the quality of images taken, and it may break down the security and control system. Privacy concerns may also come into play, because vehicle identification information could be stolen and misused. Finally, number plates may come in a variety of formats, colors, number styles, and languages (Abbas & Rashid, 2017; Jabar & Nasrudin, 2016).

In the nation of Oman, each vehicle is uniquely identified by a number plate. Number plate registration began in 1970, and the current system was initiated in 2001. Different types of number plates are differentiated by color: yellow for private vehicles, red for commercially used vehicles, white for official vehicles, and so on. The plates have different width-to-length ratios and are divided into three sections as shown in Fig. 8.1. The different sections contain Hindu-Arabic (0–9) numbers, Arabic letters, English letters, and the word "Oman" in Arabic. Figure 8.2 describes the structural organization of Oman number plates.

A few limited implementations of AVPR have already been installed in Oman, especially for parking management. For example, the International Airport of Muscat has an access control system to manage parking of vehicles. The system includes 46 cameras covering all entries and exits, all equipped with optical character recognition (OCR) technology for reading number plates. The system also provides surveillance and helps drivers to avoid traffic (Improving access control at Muscat International airport, n.d.). Another innovative parking guidance system may be found at the City Center Muscat Mall. The system helps customers to find empty parking spaces and uses LED signals which turn green or red depending on whether that area has empty spaces. In addition, the system enables people find their cars after shopping: customers can enter their vehicles number plate on the touchscreen to find where their vehicle is parked (City Centre Muscat launches new parking system, n.d.).

On the other hand, AVPR has yet to be applied to traffic control in the Sultanate of Oman. Currently, speed radar devices are still monitored by human resources in police stations who read plate numbers from camera images to enter into computer systems to retrieve vehicle records. AVPR can greatly improve the efficiency and accuracy of traffic control, hence the need for this current study.

The main objectives of this chapter may be summarized as follows:

1. To perform a detailed technical comparison between different methodologies and algorithms that are used for numbers plate recognition
2. To evaluate and test the performance of algorithms
3. To recommend the most suitable algorithm for an ANPR system in Oman

8.2 Literature Review

In this section, previous studies and research papers of number plate recognition system methodologies and algorithms are reviewed. This review involves automatic number plate recognition system for both Arabic and English-speaking countries.

ANPR systems have evolved along with the evolution of computers. The first object detection method was developed in 1963. Two- and three-dimensional image modeling was further developed in the 1970s. In 1980s mathematical algorithms for detection including edge detection and pattern recognition were discovered. By the 1990s, machine learning technologies was brought to bear in large-scale detection and recognition tasks (Fu et al., 2019). Building on these advances, ANPR systems have been developed and implemented in the USA, UK, and several European countries. In Arabic-speaking countries, practical studies using real number plates have been conducted in Saudi Arabia, Libya, Morocco, Tunisia, Iraq, and Sudan. In the following, we first review studies conducted in Arabic-speaking countries; then studies that are focused on other countries; and finally, general studies that are not country-specific.

8.2.1 Studies in Arabic-Speaking Countries

Abbas and Rashid in (Abbas & Rashid, 2017) propose a system for detecting license plates in Saudi Arabia. Their system faced technical challenges in that Saudi plates are white, as are most vehicles. Their system has four phases: image preprocessing; edge detection; object analysis; and artificial neural network (ANN) classification. Image preprocessing and morphological operations were used to improve the image, while the inputs to the ANN classifier were discrete wavelet transformation (DWT) coefficients. The ANN output was a decision on whether or not the identified object was a license plate. The optimized system was able to achieve a detection rate of 99%.

Omran and Jarallah in (Omran & Jarallah, 2018) discuss automatic extraction methods for Iraqi number plates from images by using two different methods: morphological operation and edge detection. They ran experiments using 60 vehicle images from different types and conditions. The average run time using morphological operation is 3 seconds and also works on low-resolution images but with longer runtime. The number plate extraction rate for morphological operation is 98%. The average run time for the edge detection method is 2 seconds, and the extraction percentage rate was 82%.

Mubarak et al. in (Mubarak et al., 2017) discussed algorithms of automatic vehicle number plate recognition to identify Sudanese license plates. The proposed system in this study has five stages: image capture and input, image preprocessing, vehicle plate detection, character segmentation, and character recognition. On a test set of 33 images of Sudanese vehicle number plates, the accuracy rate of detecting and locating the plate number from vehicle images was 96%, while the accuracy rate of the character extraction stage was 90%.

Taki and El-Alaoui in (Zahra Taki & El Belrhiti El Alaoui, 2018) present a system design for recognition of Moroccan number plates. The proposed system in this study has three main levels: Number plate localization is accomplished based on a hybrid method that combines edge extraction and morphological operations; number plate segmentation is based on number plate features; and character recognition level is accomplished using the Tesseract open source OCR engine (https://github.com/tesseract-ocr). This methodology is able to recognize Arabic characters.

Damak et al. in (Damak et al., 2020) present deep learning techniques of number plate recognition systems to detect and identify Tunisian number plates. The proposed system is based on algorithms for number plate localization, character segmentation, and OCR, respectively. The first two algorithms rely on contour detection and selection, while OCR is accomplished using a convolutional neural network (CNN). All these algorithms are implemented in Python. The CNN achieved an accuracy of 95.84% on the MNIST dataset, which compares favorably with previous studies.

8.2.2 Studies in Other Countries

Kilic and Aydin in (Kilic & Aydin, 2019) discuss deep learning algorithms of number plate recognition for Turkish vehicles. 4693 images of vehicles were captured from security cameras at Firat University. The preprocessing algorithm has three stages. First the image is uploaded to the system and cropped to improve performance; second, the image is converted to grayscale and then to binary image format; and third, morphology operations are performed. The images are divided into three groups: 75% for training, 20% for testing, and 5% for validation. A convolutional neural network (CNN) is used to perform classification. The algorithm is able to achieve 96.36% overall number plate identification accuracy rate, while the per-character accuracy rates were 99.43% for numbers, 99.05% for letters, and 99.31% overall.

Hidayatullah et al. in (Hidayatullah et al., 2016) present an algorithm for detection and identification of vehicles number plates for Indonesian vehicles. Their method includes two main phases: number plate localization and character recognition. Localization is accomplished using histogram-based horizontal and vertical edge detection applied to the grayscale image. The extracted license plate is converted to black and white using an Otsu threshold. Template matching was using in recognition, which is an OCR technique that is used to match the characters images with stored characters and texts. A sample of 80 Indonesian number plate images taken at a distance of 1 m and height 55 cm was used to verify the method. 78.65% of sample successfully detected the number plate. For recognition phase using successfully detected sample and 85.9% of detected sample achieved average similarity based on Jaccard similarity index, while Tesseract achieved 83.45% and original template matching achieved 49.49%.

Mondal et al. in (Mondal et al., 2017) use a four-layer CNN that works on segmented Indian number plate images and is able to recognize the state of origin of the plate. Based on a dataset of 200 images from each of four states that was divided into 30% training and 70% testing, the algorithm was able to correctly identify states with an overall accuracy of greater than 97%.

Babu and Raghunadh in (Babu & Raghunadh, 2016) discuss an algorithm for Indian number plate character recognition based on bounding box character segmentation methodology. This method has four stages: preprocessing of captured image, number plate localization, character segmentation, and character recognition. Character segmentation plays a significant role in this algorithm because a bounding box is used to segment the number plate image into different characters. A sample of 45 vehicle images was used to check performance. The localization, character segmentation, and character recognition stages had accuracies of 93.33%, 86.67%, and 93.33%, respectively. Factors that decrease the accuracy rate include image blurring, broken plates, and similarities between some characters such as O and 0, S and 5, B and 8, and so on.

Islam et al. in (Islam et al., 2015) propose an ANPR system built on morphological operations. This system uses an optical character recognition device to read characters from vehicle image. The algorithm is divided into several levels: image capture, image preprocessing, edge detection, vehicle number plate extraction, character segmentation, character matching, number plate identification, and output. The output includes identified number plate in pixels and recognized characters of number plate with template matching. For testing, they used 150 images of Bangladesh plates of size 640 × 480-pixel captured from 4 to 5 meters' distance using a high-resolution camera. The license plate number recognition rate was about 92%, and the computation time was about 0.3 seconds. Characters that are rotated up to 45° may be identified. Abnormal number plate shape, size, and longer capture distances are the primary causes of failure in this algorithm.

The ANPR algorithm of Wibirama in (Wibirama & Nugroho, 2017) has three stages: number plate detection, character segmentation, and character recognition. For plate detection, the authors used features based on zonal densities as inputs to a support vector machine (SVM) classifier. Their dataset included seven images at each distance of 1, 3, and 5 meters. The accuracy rate of character recognition is 89.8%, 82.9%, and 65.2% for the three distances, respectively.

Alhussein et al., in (Alhussein et al., 2019), discusses character segmentation and perspective correction algorithms for vehicle number plate recognition. The study uses images from different angles, weather conditions and different lighting levels. Number plate detection and slope correction are accomplished using convolution and homography, and vertical and horizontal filtering algorithms are used to determine the borders of number plate. Morphological operations and polynomial fitting are used to detect the four corner points, using 1094 vehicle images from different rotations, directions, angles, and weather conditions, using two different datasets. First set contains 1094 vehicle number plates from Indonesia that are rotated by perspective correlation algorithm, and the accuracy rate is 97.16% correctly rectified and cropped. Second set contains 150 images taken at different parking conditions, and the accuracy rate is 95.33% correctly cropped.

8.2.3 Summary of Prior Art Results

Table 8.1 summarizes materials, methods, and results for several prior art systems.

8.3 Materials and Methods

8.3.1 Overview

In this section, we describe the materials and methods required to evaluate the proposed ANPR system for Oman based on deep learning. Figure 8.4 gives an overview of the system. The figure shows three columns: number plate detection,

Table 8.1 Summary of prior art systems

Reference title	Material and methods	Results
Abbas and Rashid (2017)	Edge detection, morphological operations, and (ANN) artificial neural network classifier	Detection rate 98.68%
Omran and Jarallah (2018)	Using two methods: morphological operation and edge detection to test 60 car images	3 s run time for morphological operation and extraction percentage 98% 2 s run time for the edge detection method and extraction percentage 82%
Mubarak et al. (2017)	Using 33 images of vehicle number plates	96% of images are detecting number plate region, and character recognition rate is 90%
Damak et al. (2020)	Using CNN algorithm for number plate recognition	The accuracy rate is 95.84%
Kilic and Aydin (2019)	4693 vehicle images divided into three groups: 75% for training, 20% for testing, and 5% for validation. A convolutional neural network (CNN) is using for character recognition	96.36% overall number plate identification accuracy rate Per-character accuracy rates were 99.43% for numbers, 99.05% for letters, and 99.31% overall
Hidayatullah et al. (2016)	A sample of 80 Indonesian number plate images taken at a distance of 1 m and height 55 cm includes two main phases: number plate localization and character recognition	78.65% of sample successfully detected the number plate. For recognition phase, 85.9% achieved average similarity based on Jaccard similarity index, while Tesseract achieved 83.45% and original template matching achieved 49.49%
Islam et al. (2015)	150 images collected from different locations. This system is built into morphological operations	Recognition rate is 92%, and computation time is about 0.3 seconds, and this is able to identify the characters up to 45°
Wibirama and Nugroho (2017)	It has three stages: number plate detection, character segmentation, and character recognition 21 vehicle images dataset at 1, 3, and 5 meters distance	The accuracy rate of character recognition is 89.8%, 82.9%, and 65.2% for 1, 3, 5 meters distances
Selmi et al. (2017)	Using convolution neural network model for number plate detection and recognition	Using precision, recall and f-score rate to evaluate number plate detection More than 95% character recognition accuracy rate

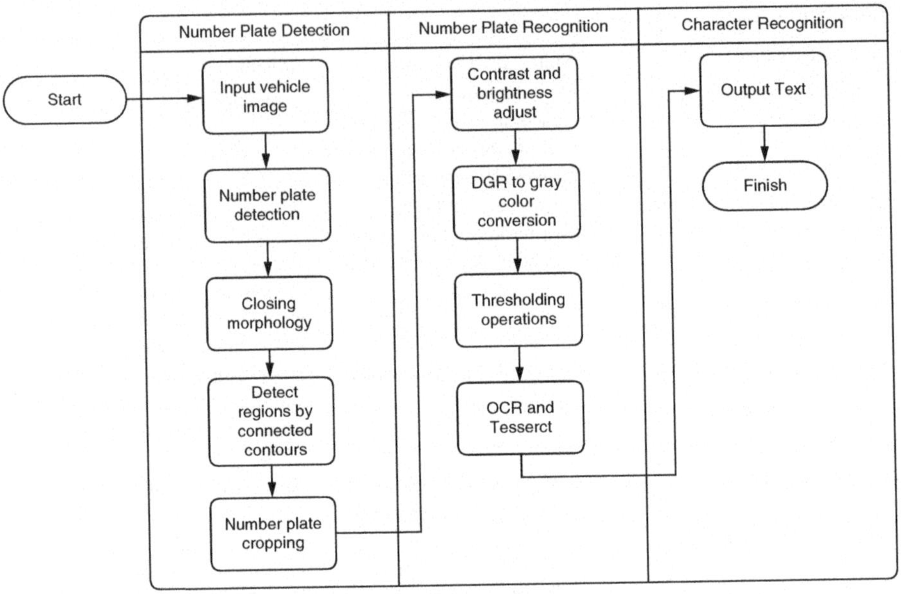

Fig. 8.3 Flowchart for recommended Omani ANPR system

number plate recognition, and character recognition. The first two columns (number plate detection and recognition) can be considered as aspects of preprocessing, which takes images of vehicles with number plates and prepares character-by-character images, while the last column (character recognition) takes these prepared images and interprets them using CNNs. In this research, the preprocessing and CNNs were evaluated separately, using separate datasets. The reason for this is that testing and training of the CNNs require much larger datasets of character images than can be obtained using Omani license plates.

This section is divided into two parts, which focus on preprocessing and deep learning, respectively. In the first part, the dataset used to assess the performance of the preprocessing algorithms is described in the first subsection. The next two subsections explain the algorithms for number plate detection and recognition, respectively, which are listed in the first two columns of Fig. 8.3. In the second part of this section, the datasets used to train and test the character recognition algorithms are described in the first subsection, followed by subsections dealing with CNN design and evaluation, respectively.

8.3.2 Number Plate Dataset

Since there was no pre-existing dataset for Omani number plates, it was thus necessary to construct a new dataset, based on images from the Internet. The dataset

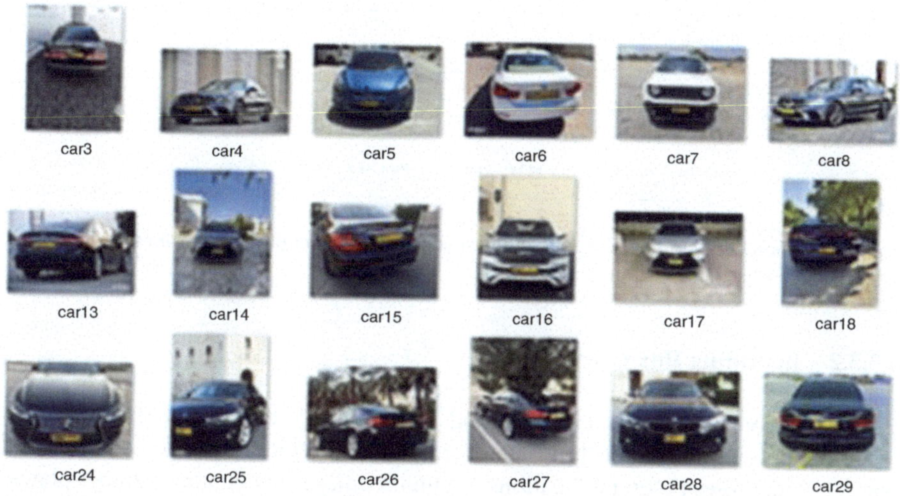

Fig. 8.4 Example images from Omani number plate dataset

contains 355 images of vehicles from different distances and capturing angles. All images are color photos in jpeg format. The camera angle of captured images was restricted to less than or equal to 45° horizontally or vertically, as recommended in (Camera Setup for Best ANPR | Plate Recognizer ALPR, n.d.). Figure 8.4 shows some vehicles with Oman's number plates from dataset.

8.3.3 Number Plate Extraction Processing Stages

Number plate detection or extraction is the first stage in the ANPR system. The main objective is to extract or isolate the region within the digital image that contains the number plate. There are many factors that affect the appearance of the number plate within the image, so preprocessing is necessary to correct for these effects. The detection phase is divided into subphases as follows:

8.3.3.1 Input Vehicle Image

The first step in this system is to insert the vehicle image to the system. Images in jpeg format are from videos or static images captured by cameras from different capturing angles. Each image has its own features: size, color density, and quality resolution that is dependent on acquisition tools and features.

```
image, contours, hierarchy= cv2.findContours(image, v2. RETR_EXTERNAL,
cv2.CHAIN_APPROX_NONE)

for i in range (0, len(contours)):
    if cv2.contourArea(contours[i]) > 200:
        x, y,w,h = cv2.boundingRect(contours[i])
        if w > 10:
            box_image = cv2.rectangle(image, (x, y), (x+w, y+h), (0, 0, 255), 1)
            cv2.imwrite(box_image)
```

Fig. 8.5 Python code fragment showing the use of OpenCV operations to form bounding boxes for possible license plates within the image

8.3.3.2 Bounding Box Creation

Edge detection is a fundamental method for feature detection or extraction. The result of applying this algorithm is an object boundary with connected curves. This operation identifies potential locations within the image that contain number plates, since a number plate is a region with a rectangular boundary. For this purpose, operations from the OpenCV library in Python are used to identify possible license plate boundaries and enclose them in bounding boxes. The code fragment in Fig. 8.5 shows the use of OpenCV operations to create bounding boxes. The code first finds contours and then constructs a rectangular bounding box around contours with enclosed area of greater than 200 pixels and width greater than 10 pixels. Subsequent processing is applied to the bounding box image.

8.3.3.3 Closing Morphological Operations

In order to properly identify the corners of the rectangular plate, the four segments of the plate boundary must be clearly delineated. In the proposed system, the boundary is regularized using morphological operations, which are a collection of a nonlinear image processing operations based on the shape of features within an image. Typically, these are applied to binary images, so the bounding box image is converted to binary (using the Otsu threshold) before morphological operations are applied. The two basic morphological operations are erosion and dilation (Jabar & Nasrudin, 2016). Both operations require two inputs: the original image to be transformed and a structuring element or kernel that decides the nature of operation. The combination of dilation followed by erosion is called closing: it fills holes within contours and fills out the boundary.

In OpenCV, morphological operations are performed using the `morphologyEx` command. In the code fragment in Fig. 8.6, a closing operation using a 5×5 square kernel is performed on binary image data contained in the array `image`.

```
kernel = np.ones((5, 5), np.uint8)
closing = cv2.morphologyEx(image, cv2.MORPH_CLOSE, kernel)
```

Fig. 8.6 Python code fragment for morphological closing of image

Fig. 8.7 Image with identified rectangular license plate contours outlined in red and labeled with "LP"

8.3.3.4 Number Plate Rectangle Detection

The result of closing is passed once more to contour detection, but this time the corners of the contour are detected. This is accomplished by using the `CHAIN_APPROX_SIMPLE` option of `findContours`, which generates contour corners. Hence the rectangular number plate can be recognized as a contour with four corners (Herusutopo et al., 2012). The specific commands in OpenCV are:

```
contours= cv2.findContours(closing, cv2.RETR_TREE,
cv2.CHAIN_APPROX_SIMPLE)
```

the `CHAIN_APPROX_SIMPLE` option returns just the corner points of the contour, which for the number plate will be the four corners of the rectangular plate (Fig. 8.7).

8.3.3.5 Number Plate Cropping

This is the final step in number plate detection stage. Cropping the number plate region as a separate image facilitates character recognition by limiting the scope (Fig. 8.8).

8.3.4 Number Plate Recognition

It is the next level of preprocessing; the extracted number plate is converted to a binary image. Doing this efficiently requires several preprocessing steps, including brightness and contrast adjustments, conversion of RGB image grayscale, and

Fig. 8.8 Number plate region cropped from the vehicle image

Original number plate	Number plate after brightness and contrast adjustments

Fig. 8.9 Contrast adjustment for license plate image (**a**) Original number plate (**b**) Number plate after brightness and contrast adjustments

thresholding to convert the grayscale image to a binary image. These steps are described in sequence below.

8.3.4.1 Brightness and Contrast Adjustments

Vehicle images taken in different bad conditions like rainy weather, fog, dark nights, or bright days. The overall image looks dim that needs to adjust the brightness and contrast to avoid any shadows or blurs that effect badly in color or region edge detection to detect ROI correctly. This adjustment is implemented using the equalizeHist command of OpenCV as follows (Fig. 8.9):

```
image_adjusted = cv2.equalizeHist(image)
```

8.3.4.2 RGB to Grayscale Conversation

Colored images include R, G, and B channels, where each channel has 256 levels which are specified by 8 bits. The following standard equation is used to convert R,G,B, triples to an 8-bit grayscale image (Fu et al., 2019; Ghadage & Khedkar, 2020):

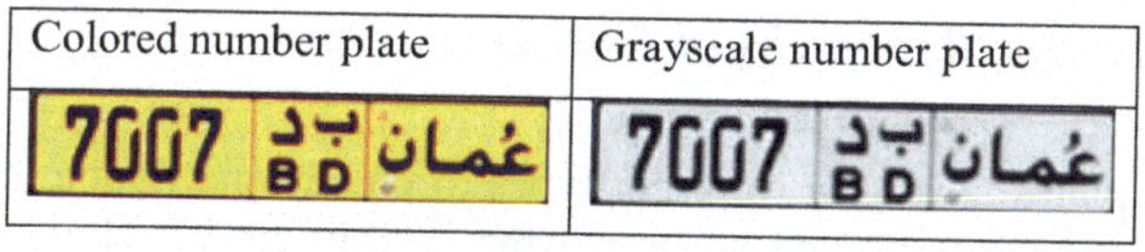

Fig. 8.10 Grayscale conversion of number plate image. (**a**) Colored number plate. (**b**) Grayscale number plate

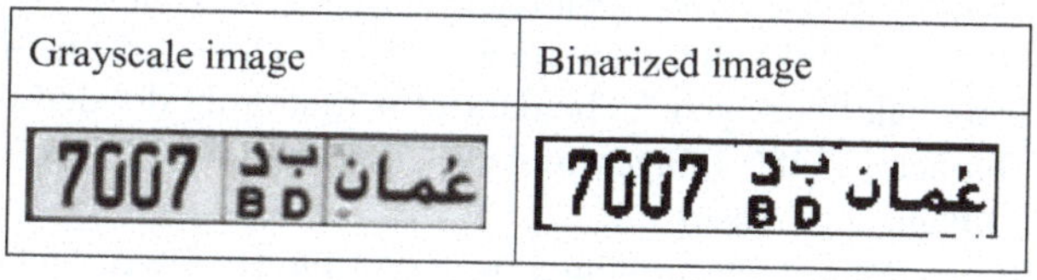

Fig. 8.11 Thresholding and binarized image. (**a**) Grayscale image. (**b**) Binarized image

$$Grayscalevalue = 0.299R + 0.587G + 0.114B$$

This conversion is implemented in the cvtColor command of OpenCV, using the COLOR_BGR2GRAY option as follows (Fig. 8.10):

```
gray = cv2.cvtColor(image_adjusted, cv2.COLOR_BGR2GRAY)
```

8.3.4.3 Image Binarization

Optical character recognition algorithms typically work on binary images, because gray levels do not give additional meaningful information. The conversion is affected by choosing a suitable threshold and mapping all grayscale pixel values that exceed the threshold to 1 and all other pixels to 0. A common choice of threshold is the Otsu threshold, which maximizes the between-class variance while minimizing within-class variances. In OpenCV, this is implemented using the `threshold` command using the THRESH_OTSU option (Fig. 8.11).

```
ret, th = cv2.threshold(gray, 0, 255, cv2.THRESH_BINARY
+ cv2.THRESH_OTSU).
```

8.3.5 Character Segmentation

The segmented number plate image must be further segmented into rectangles containing individual characters. There are a number of available character segmentation algorithms, as described in Sect. 8.2. Since character segmentation is a straightforward process, this stage was not evaluated in the current research.

8.3.6 *Character Recognition Datasets*

Oman license plates have four different types of characters: Arabic letters, Arabic digits, capital Latin letters, and western Arabic digits (0–9). Thus, four different training/testing sets are required, which are used to obtain four different sets of CNN parameters. These datasets are described as follows.

The Arabic letters dataset was obtained from (Arabic Handwritten Characters Dataset, n.d.). It has 16,800-character images divided into 28 classes for the 28 Arabic letters from “alef” to “yeh.” The dataset is randomly divided into a training set of 13,440 characters (480 images in each class) and a testing dataset of 3360 characters (120 images per class).

The Arabic digits dataset was obtained from (Arabic Handwritten Digits Dataset, n.d.). It has 60,000 training images and 10,000 testing images. Both datasets are evenly divided into ten classes from siffr (zero) to tissaa (nine). The Latin alphabet dataset was obtained from (English Alphabets Dataset, n.d.). It contains 26 classes for the capital letters from A to Z. The dataset contains 372,450 handwritten character images in grayscale. The dataset is divided into 80% training and 20% testing. The Hindu-Arabic digit dataset was obtained from (Digit Recognizer, n.d.). It contains 42,000 training images and 28,000 testing images. Both sets are equally divided into 10 classes from 0 to 9.

All images from the different testing and training sets are converted to binary using the Otsu threshold, and are scaled to size 32×32. These images then serve as inputs to the CNN.

8.3.7 *Character Recognition Algorithms*

8.3.7.1 Overview of Character Recognition Subsystem

Although several optical character recognition (OCR) algorithms have been developed over the years, in recent years CNN-based algorithms have surpassed all others. CNNs are particularly capable of identifying local features in images and have greatly reduced parameters compared to fully connected neural networks.

Our model is created using the Python Keras library, which is built on the TensorFlow machine learning platform. In addition to Keras, Python has several libraries that can be used for machine learning, including NumPy (for numerical mathematics), Pandas (for data manipulation), CSV (for data input/output), and PIL (for image file manipulation).

CNN Model Architecture

Figure 8.12 shows the layer structure of the CNN model. Layers include convolutional, batch normalization, max pooling, and dropout layers one after the other.

```
model = create_model()
model.summary()

Model: "sequential"
_________________________________________________________________
Layer (type)                 Output Shape              Param #
=================================================================
conv2d (Conv2D)              (None, 32, 32, 16)        160
_________________________________________________________________
batch_normalization (BatchNo (None, 32, 32, 16)        64
_________________________________________________________________
max_pooling2d (MaxPooling2D) (None, 16, 16, 16)        0
_________________________________________________________________
dropout (Dropout)            (None, 16, 16, 16)        0
_________________________________________________________________
conv2d_1 (Conv2D)            (None, 16, 16, 32)        4640
_________________________________________________________________
batch_normalization_1 (Batch (None, 16, 16, 32)        128
_________________________________________________________________
max_pooling2d_1 (MaxPooling2 (None, 8, 8, 32)          0
_________________________________________________________________
dropout_1 (Dropout)          (None, 8, 8, 32)          0
_________________________________________________________________
conv2d_2 (Conv2D)            (None, 8, 8, 64)          18496
_________________________________________________________________
batch_normalization_2 (Batch (None, 8, 8, 64)          256
_________________________________________________________________
max_pooling2d_2 (MaxPooling2 (None, 4, 4, 64)          0
_________________________________________________________________
dropout_2 (Dropout)          (None, 4, 4, 64)          0
_________________________________________________________________
conv2d_3 (Conv2D)            (None, 4, 4, 128)         73856
_________________________________________________________________
batch_normalization_3 (Batch (None, 4, 4, 128)         512
_________________________________________________________________
max_pooling2d_3 (MaxPooling2 (None, 2, 2, 128)         0
_________________________________________________________________
dropout_3 (Dropout)          (None, 2, 2, 128)         0
_________________________________________________________________
global_average_pooling2d (Gl (None, 128)               0
_________________________________________________________________
dense (Dense)                (None, 28)                3612
=================================================================
Total params: 101,724
Trainable params: 101,244
Non-trainable params: 480
```

Fig. 8.12 Keras output showing CNN model structure

This sequence of four layers is repeated four times. The purpose of each type of layers is as follows.

- Convolutional layers produce feature maps that retain local features within the image. Each feature map in each convolutional layer uses a 3 × 3 kernels (which is optimized during the training process), and rectified linear unit (ReLU) activation functions are applied to convolution outputs. The four successive convolutional layers have 16, 32, 64, and 128 feature maps, respectively.
- Following each convolutional layer is a batch normalization layer. In the course of training, the distribution of convolutional layer outputs shifts over time, which leads to slower learning rates for subsequent layers. Batch normalization addresses this problem by standardizing the distribution of outputs from convolutional layers, thus making it possible to accelerate the learning for subsequent layers.
- After the batch normalization layers are max pooling layers. Max pooling layers retain important features and reduce the number of input layers to the next layer, thus reducing the number of parameters and decreasing training time.
- Max pooling layers are followed by dropout layers, which like batch normalization is only used during training. For each training instance, the dropout layers thin out the information passed to the next layer by randomly setting to 0 a fixed percentage of the layer inputs. As a result, the subsequent layer become less dependent on individual neurons or fixed patterns of neurons, thus making it more robust. In our case, the percentage used was 20%.
- Global average pooling is used to replace fully connected layers to create one feature map for each corresponding types of the classification.
- Dense layer is a deeply connected neural network each neuron in dense layer receives input from all neurons of its previous layer.
- The final layer has the same number of outputs as there are classes to be identified. The softmax function is used to convert the signed outputs into positive numbers that sum to 1. These numbers are interpreted as probabilities of the different classes.

8.3.7.2 CNN Training and Evaluation

Training of the CNN requires an accuracy measure in order to guide the training process. Since the system is for multi-class classification, categorical cross-entropy is an appropriate loss function to use and accuracy as metrics.

The training process uses stochastic gradient descent and backpropagation. The process is governed by the learning rate parameter, which determines the size of incremental changes to the model parameters based on each training instance. There are various algorithms for dynamically adjusting the learning rate. One of the most popular is the Adam optimizer (Alto, n.d.). The actual command for implementing the training process in Keras is as follows:

```
model=create_model (optimizer='Adam', kernel_initializer='
uniform', activation='relu')
```

8.4 Results and Discussion

8.4.1 Outline of This Section

In this research, the license plate extraction and character identification portions of the system were tested separately, and the results are described in the two following subsections, respectively:

8.4.2 Number Plate Detection and Recognition Results

8.4.2.1 Image Processing Outcomes

Table 8.2 gives examples of outcomes from different processing steps in number plate detection and recognition. The results are collecting from different angles, different image resolution, and different distances.

8.4.2.2 Detection and Recognition Overall Accuracies

The overall accuracies of different stages in number plate detection and segmentation are summarized in Table 8.3.

8.4.2.3 Factors that Affect Number Plate Detection and Recognition

Figure 8.13 shows examples of detection errors in number plate detection. In some cases, only a part of the license plate is captured; in others, the characters are vague and incomplete. Images were manually categorized into several categories according to the principal image characteristics. Image categories were rear or front angle (for images that were directly from front or rear); left or right angle (for images that were taken from the side); top angle (for images that were taken from an elevated position); close distance (for images that were taken from short distance length between vehicle and camera); wide or long distance (for images that were taken from long or wide distance); and night lighting (for images that were taken at nights with artificial lighting, car or road lighting not normal day lighting).

In order to determine the effect of various factors (such as direction, distance, and lighting) on the accuracy of number plate recognition, images in the database were categorized manually according to their most prominent features, and the recognition accuracy percentage in each category was computed.

Categories included rear or front angle (i.e., the photo was taken directly from the front or back, at a moderate distance in daytime lighting); left or right angle (i.e., the photo was taken at moderate distance in daytime, but toward the left

Table 8.2 Examples of number plate images at different preprocessing stages

	Preprocessing steps	Examples of images resulting from each step		
1	Input the vehicle image			
2	Cropping number plate with rectangle curve			
3	Number plate or edge detection			
4	Morphological operations			
5	Detected number plate			
6	Brightness and contrast adjustments			
7	RGB to gray conversion			
8	Thresholding and binarized image			

or right side); top angle (i.e., the camera was significantly above the level of the plate); close distance (i.e., the camera was closer than normal from the vehicle); wide/long distance (i.e., the camera was far from the vehicle, and the vehicle and plate occupied only a part of the image); and night lighting. Figure 8.14 shows the result of this investigation. Accuracy percentage of correct detection is between 80% and 53% according to each factor sample. Night lighting caused the greatest reduction in accuracy. Interestingly, left/right and top angle actually had better accuracy than direct front/back images: note however that according to the

Table 8.3 Accuracy rates for different processing stages in number plate detection and segmentation

Processing stage	# of input images	# Correct	Accuracy (%)
Detection of number plate	130	93	71.5%
Formation of bounding box containing number plate	93	88	94.6%
Identification of number plate rectangular boundary	88	83	94.3%
Combined accuracy (detection + bounding box + identification)	**130**	**83**	**63.8**

Fig. 8.13 Examples of license plate detection errors

selection criteria, all images had angles less than 45° horizontally and vertically, as recommended in (Camera Setup for Best ANPR | Plate Recognizer ALPR, n.d.) (note however that others recommend a maximum angle of 30° (Angle of Capture | Rekor Documentation, n.d.)). Wide and long distance was 13.3% less accurate than close distance, which reflects the fact that the size of the number plate in pixels is important: this is because the number of pixels per plate determines the resolution. Other factors that can influence accuracy are weather conditions such as rain or fog: these were not included as categories due to lack of images (Fig. 8.14).

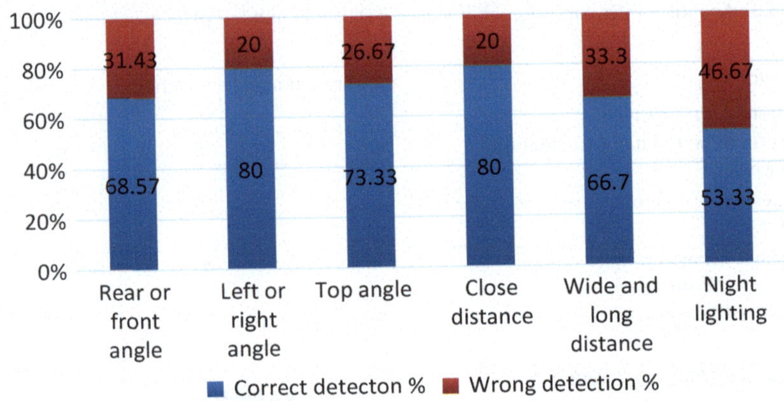

Fig. 8.14 Detection rate according to license plate image conditions

8.4.2.4 Average Processing Time

The time for recognition processing was computed for 29 vehicles. Values ranged between 0.25 and 0.45 seconds, with an average 0.31 seconds. These results confirm the system has an acceptable run time for real-time application.

8.4.3 Results of Using CNN Algorithm for Character Recognition

8.4.3.1 Training the CNN Model

Separate models were trained for each of the four kinds of characters. The lengths of the training processes are characterized in terms of the number of training epochs, which is the number of times the entire training dataset is passed through the model. All models were trained twice, with two different training lengths, in order to verify convergence of the model. Part of each testing set was reserved for validation: for all four datasets, the validation set size was equal to the testing set size and consisted of characters evenly divided among classes. Typically, the training and validation losses decrease and accuracies increase as the training progresses until the model has converged. Usually training loss starts out higher than validation loss, and when training loss falls below validation loss, it is taken as an indication that overfitting is taking place.

Figure 8.15 shows sparse cross entropy loss and classification accuracy for training and validation phases as a function of epoch, for Arabic letters (left) and digits (right). Both models were trained twice with batch size 20, using 10 and 20 epochs, respectively. As expected, the plots show generally decreasing loss and increasing accuracy as a function of epoch. After 10 epochs, the training loss for

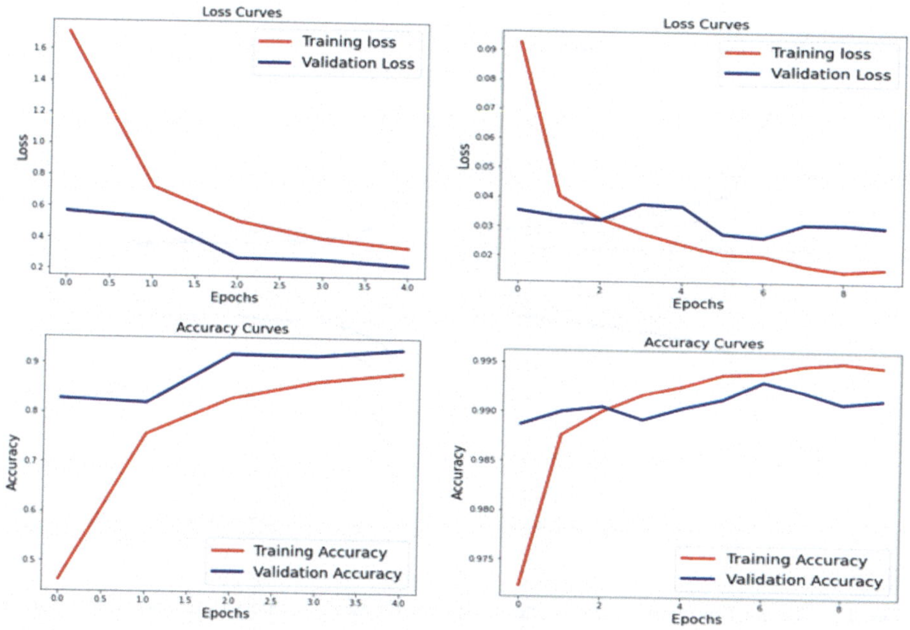

Fig. 8.15 Training and validation losses and accuracies for Arabic letters (left) and digits (right) datasets

the digit dataset has fallen below the validation loss, indicating that the model has converged and further training may lead to overfitting. Indeed, training the model over 20 epochs yielded no significant improvement in validation accuracy. In contrast, the Arabic letters model has not yet converged within ten epochs. When trained again with 20 epochs, the Arabic letters test accuracy increases to 96.14%, compared to 94.1% on 10 epochs.

For the English letters model, a batch size of 32 was used for 5 and 30 epochs, while for Hindu-Arabic digits the batch size was 64 with 10 and 15 epochs. Figure 8.16 shows training and validation accuracies on 5 epochs of English letters and 10 epochs of Hindu-Arabic numbers. The test accuracy is increased from 99.31% on 5 epochs to 99.77% on 30 epochs in letters dataset, while the test accuracy is increased from 98.31% on 10 epochs to 98.85% on 15 epochs in numbers dataset.

8.4.3.2 Character Recognition Models' Performance

Besides cross-entropy loss and accuracy, there are several common metrics used to evaluate the performance of prediction models. Three important measures of accuracy are precision, recall, and F1-score. Precision reflects the rate of false positives; recall indicates the frequency of false negatives; and F1 score balances the two types of errors. Since all classes are balanced in this case, average recall will be

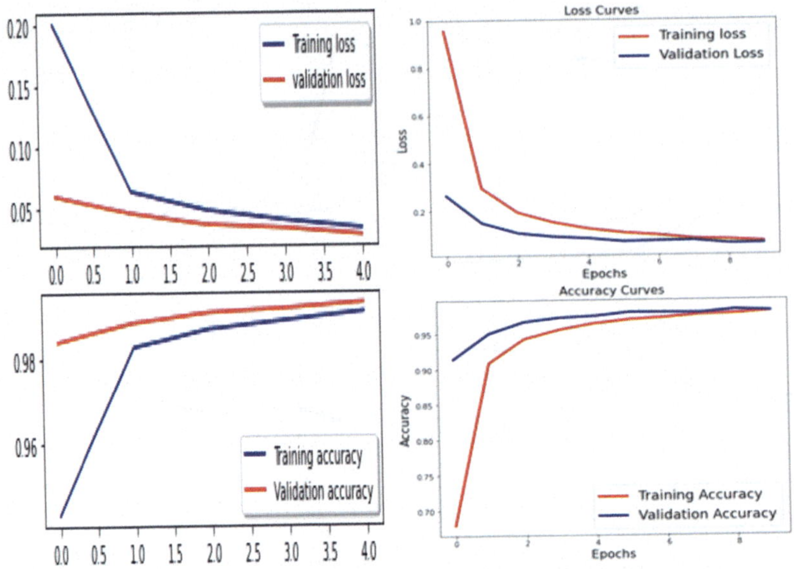

Fig. 8.16 Training and testing losses and accuracies for English letters (left) and Hindu-Arabic digits (right)

Table 8.4 Precision, recall, and f1 score for four datasets

Dataset name	Average precision	Average recall	Average f1-score
Arabic characters (28 classes)	96%	96%	96%
Arabic digits (10 classes)	99%	99%	99%
Latin characters (26 classes)	99%	99%	99%
Hindu-Arabic numbers (10 classes)	100%	100%	100%

the same as average accuracy. Table 8.4 compares these three-accuracy metrics for the four different datasets used in the system. Hindu-Arabic numbers achieve the highest accuracies, while Arabic characters obtain the lowest scores. This is to be expected, because Arabic characters have the most classes to distinguish.

8.5 Conclusions

This chapter has thoroughly described and verified a novel system for extracting and recognizing English and Arabic characters and digits from Oman's vehicles number plates. The system is based on three key image processing algorithms: morphological operations for number plate detection; thresholding operations for number plate recognition; and CNN for character recognition.

Several datasets were used to evaluate the accuracy and execution time of algorithms in the system. The execution time for recognition was estimated at between 0.25 and 0.45 seconds, which is sufficiently short to make the system suitable for real-time operation in realistic traffic conditions. Additionally, number plate overall extraction accuracy was 71.5%, while the CNN modeling gave between 96% and 100% accuracy on extracted characters.

For future work the algorithms may be fine-tuned, particularly the number plate recognition which was the least accurate component of the system. Character extraction may also be added, and then the complete system may be practically implemented, possibly through the use of mobile smart devices.

References

Abbas, A. M., & Rashid, A. E. (2017). Saudi Arabia license plate detection based on ANN and objects analysis. *International Journal of Applied Engineering Research ISSN, 12*(13), 3740–3751.

Alhussein, M., Aurangzeb, K., & Haider, S. I. (2019). Vehicle license plate detection and perspective rectification. *Elektronika ir Elektrotechnika*, ISSN 1392-1215., *25*(5), 47–57.

Alto, V. (n.d.). Deep learning for image recognition: Convolutional Neural Network with Tensorflow and Keras. [Online]. Available: https://towardsdatascience.com/deep-learning-for-image-recognition-convolutional-neural-network-with-tensorflow-de6349c31c07. Accessed 9 Mar 2021.

Angle of Capture | Rekor Documentation. (n.d.). [Online]. Available: https://docs.rekor.ai/camera-configuration/camera-placement-guide/angle-of-capture. Accessed 9 Mar 2021.

Arabic Handwritten Characters Dataset. (n.d.) *Kaggle*. [Online]. Available: https://www.kaggle.com/mloey1/ahcd1. Accessed 9 Mar 2021.

Arabic Handwritten Digits Dataset. (n.d.). *Kaggle*. [Online]. Available: https://www.kaggle.com/mloey1/ahdd1. Accessed 9 Mar 2021.

Babu, M., & Raghunadh, M. V. (2016). Vehicle number plate detection and recognition using bounding box method. In *2016 international conference on advanced communication control and computing technologies* (Vol. 978, pp. 106–110). https://doi.org/10.1109/ICACCCT.2016.7831610

"Camera Setup for Best ANPR | Plate Recognizer ALPR." (n.d.) [Online]. Available: https://platerecognizer.com/camera-setup-for-best-anpr/. Accessed 8 Apr 2021.

"City Centre Muscat launches new parking system." (n.d.) [Online]. Available: http://www.tradearabia.com/news/RET_309514.html.

Damak, T., Kriaa, O., Baccar, A., Ben Ayed, M. A., & Masmoudi, N. (2020). Automatic number plate recognition system based on deep learning. *World Academy of Science, Engineering and Technology, International Journal of Computer and Information Engineering, 14*(3), 1–6.

Digit Recognizer. (n.d.). *Kaggle*. [Online]. Available: https://www.kaggle.com/c/digit-recognizer/data. Accessed 9 Mar 2021.

English Alphabets Dataset. (n.d.). *Kaggle*. [Online]. Available: https://www.kaggle.com/sachinpatel21/az-handwritten-alphabets-in-csv-format. Accessed 9 Mar 2021.

Fu, U., Hua, Y., Ren, S., & Oymak, S. (2019). *Automatic license plate recognition using neural network and signal processing*. University of California Riverside.

Ghadage, S. S., & Khedkar, S. R. (2020). Automatic number plate recognition system using Raspberry Pi. *International Journal of Innovative Technology and Exploring Engineering, 9*(2), 1863–1865. https://doi.org/10.35940/ijitee.b7919.129219

Ghasempour, S. (2015). *Automatic License Plate Recognition (ALPR)*. Eastern Mediterranean University.

Herusutopo, A., Zuhrudin, R., Wijaya, W., & Musiko, Y. (2012). Recognition design of license plate and car type using tesseract OCR and EmguCV. *CommIT (Communication and Information Technology) Journal, 6*(2), 76. https://doi.org/10.21512/commit.v6i2.573

Hidayatullah, P., Feirizal, F., Permana, H., Mauluddiah, Q., & Dwitama, A. (2016). License plate detection and recognition for Indonesian cars. *International Journal on Electrical Engineering and Informatics, 8*(2). https://doi.org/10.15676/ijeei.2016.8.2.7

"Improving access control at Muscat International airport." (n.d.) [Online]. Available: https://www.quercus-technologies.com/news/reference/improving-access-control-at-muscat-international-airport. Accessed: 7 May 2021.

Islam, R., Sharif, K. F., & Biswas, S. (2015, December). Automatic vehicle number plate recognition using structured elements. In *2015 IEEE conference on system, process & control* (pp. 44–48). https://doi.org/10.1109/SPC.2015.7473557.

Jabar, K. A., & Nasrudin, M. F. (2016). Libyan vehicle plate recognition using region based features and probabilistic neural network. *Journal of Theoretical and Applied Information Technology, 94*(1), 104–114.

Kilic, I., & Aydin, G. (2019, April). Turkish vehicle license plate recognition using deep learning. In *2018 international conference on artificial intelligence and data processing IDAP 2018* (pp. 1–5). https://doi.org/10.1109/IDAP.2018.8620744.

Mondal, M., Mondal, P., & Saha, N. (2017). Automatic number plate recognition using CNN based self synthesized feature learning. In *2017 IEEE Calcutta conference* (pp. 378–381).

Mubarak, H., Ibrahim, A. O., Elwasila, A., & Bushra, S. (2017) Sudanese license plate identification using automatic number plate recognition. In *2017 joint international conference on information and communication technologies for education and training and international conference on computing in Arabic.*

Omran, S. S., & Jarallah, J. A. (2018). Automatic Iraqi cars number plates extraction. *Iraqi Journal for Computers and Informatics by Univ. Inf. Technol. Commun. Autom., 44*(1), 23–30.

Saghaei, H. (2016, October). Proposal for automatic license and number plate recognition system for vehicle identification.

Selmi, Z., et al. (2017). Deep learning system for automatic license plate detection and recognition. https://doi.org/10.1109/ICDAR.2017.187

Singh, I. (2016). *Automatic vehicle detection and recognition*. University of Windsor.

Wibirama, S., & Nugroho, H. A. (2017). Long distance automatic number plate recognition under perspective distortion using zonal density and support vector machine. In *2017 3rd international conference on computation for science and technology* (pp. 1–6).

Zahra Taki, F., & El Belrhiti El Alaoui, A. (2018, March). Moroccan License Plate recognition using a hybrid method and license plate features. *ResearchGate*.

Chapter 9
Real-Time Detection of First Stories in Twitter Using a FastText Model

Samar Elbedwehy, Christopher Thron, Mohammed Alrahmawy, and Taher Hamza

9.1 Introduction

The Internet revolution has enabled everyone with an Internet connection to broadcast information to a wide audience. Concurrently, the rise in mobile device usage along with many applications available for social networks allows users to publish updates easily and frequently. The result has been a huge and constantly growing volume of information. A good example of this information explosion is Twitter, which is one of the most popular social networks used worldwide. Twitter has 310 million monthly active users and one billion unique visits monthly to sites with embedded tweets (Twitter formal website, n.d.). Every day, more than 400 million new tweets are generated (McCreadie et al., 2013), including retweets of others' tweets. These tweets present an opportunity for information retrieval at web-scale for text, images, or videos associated with a particular news item.

One of the most commonly studied information retrieval tasks in the Twitter domain is event detection, where "event" here refers to an actual occurrence at a specific place and time. In particular, first story detection (FSD) refers to locating

S. Elbedwehy (✉)
Faculty of Artificial Intelligence, Kafrelsheikh University, Kafr El-Shaikh, Egypt
e-mail: samarelbedwehy@ai.kfs.edu.eg

C. Thron
Department of Science and Mathematics, Texas A&M University-Central Texas, Killeen, TX, USA
e-mail: thron@tamuct.edu

M. Alrahmawy · T. Hamza
Faculty of Computer and Information Science, Mansoura University, Mansoura, Egypt
e-mail: mrahmawy@mans.edu.eg; taher_hamza@mans.edu.eg

M. Alloghani et al. (eds.), *Artificial Intelligence for Data Science in Theory and Practice*, Studies in Computational Intelligence 1006,
https://doi.org/10.1007/978-3-030-92245-0_9

the first tweet that refers to a particular event. FSD is useful in tracing sources of misinformation or to enable contact with the original author for verification.

New event detection, also known as first story detection (FSD), is defined within the topic detection and tracking as one of the subtasks (Allan, 2002). Given a sequence of stories, the goal of FSD is to identify the first story to discuss a particular event. In this context, an event is taken to be something that happens at some specific time and place, e.g., an earthquake striking the town of L'Aquila in Italy on April 6, 2009. Detecting new events from tweets carries additional problems and benefits compared to traditional new event detection from newswire. Problems include a much higher volume of data to deal with and also a higher level of noise. A major benefit of doing new event detection from tweets is the added social component. A well-known algorithm for FSD is term frequency-inverse document frequency (TF-IDF), which is commonly used in Twitter-oriented FSD Systems. TF-IDF assigns a weight to terms (i.e., words) according to their frequency characteristics. Words that appear multiple times in a limited number of tweets are assigned a high weight: while words that appear sporadically and words that appear in lots of tweets (such as "the" in English) are given a lower weight. Several modified versions of TF-IDF exist which use different weighting schemes which take other factors into account, such as length of documents, average term frequency in documents, etc. (Petrović et al., 2010).

Due to the rapid development of neural networks and the need for efficient representation for word and sentence classification, a team of researchers from Facebook developed a library called FastText which may be used for text classification and word representation (https://research.fb.com/blog/2016/08/FastText/). FastText builds on word2vec, which is a technique for efficiently representing word associations that was previously developed by Facebook (Meyer, 2016). In word2vec, words are represented as vectors—while in FastText, the vector representation is extended to subword units. These vector representations are constructed using an algorithm based on a shallow feed forward neural net architecture. To speed up training, both word2vec and FastText employ a ML technique called hierarchical softmax which takes advantage of the unbalanced distribution of classes.

In this chapter, we compare the FSD performance of FastText with mTF-IDF and TF-IDF (Elbedwehy et al., 2017). All algorithms are implemented on Apache Storm, which is an open-source real-time distributed event processing system (https://storm.apache.org/). Storm provides parallel, scalable, fast, and fault-tolerant processing, which is suitable for real-time processing of Twitter streams.

This chapter is organized as follows. Section 9.2 reviews related work; Section 9.3 gives the required technical background; Section 9.4 explains the methodologies used in the investigation; Section 9.5 presents the results. Section 9.6 presents the conclusion and future work.

9.2 Related Work

In this section we give an overview of relevant previous research work on the subject of FSD.

FSD is often presented in the context of a streaming situation where data arrive continuously in chronological order and every new instance must be processed in bounded space and time. A typical example is Petrovic et al. (2010) that works on a streaming model of Twitter. They assign a "novelty score" to each tweet in the Twitter stream to each tweet. They also use the cosine similarity between "term frequency-inverse document frequency (TF-IDF)" weighted vector representations of tweets to analyze threads, i.e., subsets of tweets that all discuss the same topic. TF-IDF is the most popular algorithm used for similarity detection in FSD and is used in many studies (Yang et al., 1998; Petrović et al., 2010; Benhardus & Kalita, 2013; Huddar et al., 2014; Allan et al., 2000).

Other similarity measures are also used. In 2016 Vuurens (Vuurens & de Vries, 2016) proposed a method for news summarization based on 3-nearest neighbor clustering which is more effective than a baseline that uses dissimilarity of an individual document from its nearest neighbor. In 2013 and 2014, the so-called BM25 algorithm was introduced to measure similarity between a given query and documents within a given collection (Kali et al. 2013; Brigadir et al. 2014).

In 2017 a modified version TF-IDF (mTF-IDF) was introduced to improve the converting words to vectors (Sabbah et al., 2017). Researchers used locality sensitive hashing (LSH) to build FSD system for first time in (Indyk & Motwani, 1998). This approach aims at reducing the number of comparisons that are required to find nearest neighbors among a collection of tweets. More recently, authors in (Petrović et al., 2010) performed first story detection on streaming data with application on Twitter using LSH with a proposed a variance reduction strategy which the number of comparisons performed for each query.

In 2012, Vogiatzis (Vogiatzis, 2020) worked on improving the scalability and the speed of FSD, as previous approaches focused only on the accuracy. He used a distributed approach based on the Storm open-source platform, which has the computational power process a high volume of input data process them in real time.

In 2013 Mikolov (Mikolov et al., 2013) presents a method for efficient, unsupervised computation of continuous vector representations of words. His results form the basis of the open-source toolkit named Word2vec. Since then, several authors have used word2vec for FSD, resulting in improved error rates compared to previous systems; their method uses word2vec to improve in error by reach at 13% as they make comparisons to other FSD systems (Moran et al., 2016, Repp & Ramampiaro, 2016, Panagiotou et al., 2016). The authors in (Soliman et al., 2017) built six different word embedding models for the Arabic language using three different resources including Twitter. They compared the continuous bag of words (CBOW) and Skip-gram methodologies (both included within word2vec) and showed that skip-gram is better in semantic tasks than CBOW.

Word2vec doesn't support subword information and thus cannot take advantage of morphemes which give clues to the meanings of words. To take advantage of this information, subword embedding models have been developed, which explicitly incorporates morphology into character-level compositions. Many application character-level models, including entity recognition (Plank et al., 2016), part-of-speech tagging (Ballesteros et al., 2015), language modeling (Yu & Vu, 2017), and dependency parsing (Kim et al., 2016; Wieting et al., 2016). Subword embeddings form the basis of Facebook's FastText classifier (Bojanowski et al., 2017; Joulin et al., 2017; Vinhkhuc, 2016), which can train huge models using billions of examples very rapidly.

Cao and Rei in (Cao & Rei, 2016) develop a character-level embedding that they call "char2vec," which produces word embedding by using information from morphemes (such as prefixes and suffixes) as well as characters. The word embedding function is generated using a skip-gram approach (predicting neighboring words) as well as an LSTM applied to word splittings whose outputs are used to train an attention model. The model was tested on various semantic and syntactical tasks and performed relatively well (compared to word-level models) on words containing multiple morphemes.

In 2014 Le and Mikolov (Godin et al., 2015) propose a method they call Paragraph Vector or Doc2vec where the Word2vec model is adopted to create an unsupervised algorithm for learning a fixed-length feature representation from variable-length pieces of text.

In 2016 Vosoughi et al. presented *tweet2vec*, which generates vector representations of tweets that may be used in a variety of tweet categorization tasks (Vosoughi et al., 2016). Tweet vectors are produced by taking a character-level matrix representations of tweets and passing them through a convolutional neural network (CNN) consisting of several convolutional layers, followed by a long short-term memory (LSTM) encoder-decoder. The CNN-LSTM is trained on a large corpus of three million tweets, augmented using synonym replacements. The resulting network produces vector representations that can be used as features in semantical tasks. Compared to prior art systems, tweet2vec performed better on two tasks involving semantic evaluation and sentiment classification respectively.

In this chapter we implemented FastText and make comparison between the results with TF-IDF, mTF-IDF, and FastText. More detailed explanations of these concepts are given in Sect. 9.4.

9.3 Technical Background

Section 9.2 introduces several key concepts used in FSD research, including TF-IDF, mTF-IDF, word2vec, char2vec, and FastText. The system developed in this research is built on these concepts. In this section, we explain these concepts in more detail. This chapter provides an outline of the research methodology used, a

Algorithm 1: Pseudo code for the General Approach of FSD

```
Data: corpus
foreach tweet t in corpus do
    foreach word w in t do
        foreach previously seen tweet t' that contains w do
            update distance (t, t')
        end
    end
    dis(min)(t)=min_t'{distance(t,t')}
    add t to inverted index
end
```

Fig. 9.1 Standard algorithm for FSD in Twitter

description of the data collection process and the methods that most use for this task, our method in detail, and tool which help to implement the method.

9.3.1 Basic Algorithm for Real-Time First Story Detection

In order to detect a first story in Twitter, it is necessary to identify tweets that differ significantly from previous ones. For this reason, a notion of "distance" is needed to quantify differences between tweets. Using this distance, a "novelty score" can be assigned to each incoming tweet as the minimum distance between the new tweet and previously seen tweets. If the novelty score is above a threshold, then the new tweet is labeled as a first story. Fig. 9.1 shows the exact pseudo code used by the UMass system (Allan et al., 2000), where $dis_{(min)}(t)$ is the novelty score assigned to tweet t. In order to decrease the running time, tweets are represented using only n features with the highest weights. The UMass system uses an inverted index (i.e., reduced set of previous tweets for comparison with new tweets) which optimizes the system for speed and makes sure a minimal number of comparisons are made.

9.3.2 Term Frequency-Inverse Document Frequency

Many text-mining applications related to information retrieval or classification require an algorithm for determining the relative importance of a word for determining the meaning of a text in which a document appears. Term frequency inverse document frequency (TF-IDF) is the most frequently used algorithm for this purpose and is notable for its efficiency and simplicity (Ramos, 2003). TF-IDF can be seen as a product of two factors, one reflecting term frequency and the other reflecting inverse document frequency. Given a list of n terms that occur in a set of N indexed documents, the exact expression for TF-IDF for the t'th listed term within the $d'th$ document is

$$\text{TF-IDF}_{t,d} = \frac{tf_{t,d}}{\sqrt{\sum_{t'=1}^{n} tf_{t',d}^{2}}} * \log\left(\frac{N}{DF_t}\right)$$

where $tf_{t,\,d}$ is the number of occurrences of term t within document d and DF_t is the number of documents in which term t occurs.

9.3.3 *Modified Term Frequency-Inverse Document Frequency*

Modified TF-IDF (mTF-IDF) has been shown to lead to better performance in some text mining tasks (Meetei et al., 2019). The main idea in mTF-IDF algorithm is not only to consider the number of occurrence of term t in document d, but also it takes in consideration the proportion of the total number of occurrences of t in all documents (denoted by T_t)as well as the total number of terms in the entire collection of documents (denoted by T_c). The actual formula is

$$\text{mTF-IDF}_{t,d} = \frac{tf_{t,d}\cdot\left(\frac{T_t}{T_c}\right)}{\log\left[\left(\sum_{t=1}^{n} tf_{t,d}^{2}\right)\cdot\left(\frac{\text{length}_d}{T_c}\right)\right]}\cdot\log\left(\frac{N}{\text{DF}_t}\right)$$

A comparison between TF-IDF and mTF-IDF is given in (Elbedwehy et al., 2017). In the current system, both are implemented on the Storm platform described in Sect. 9.3.6.

9.3.4 *Word Embeddings Using Word2vec*

9.3.4.1 Overview

Word embeddings refer to mappings from words to numerical representations, often in the form of vectors. A simple example of word embedding is one-hot encoding, where each word is represented as a vector with a single "1" entry and all other entries equal to zero. This requires a vector space whose dimension is equal to the total number of words in the dataset. Furthermore, one-hot encoding does not capture information about a word's meaning or context—hence potential relationships, such as contextual closeness, are not captured across collections of words. For example, one-hot encodings of "dog" and "cat" do not reflect the fact both refer to animals. Nonetheless, such encodings can provide sufficient baselines for simple NLP tasks (e.g., email spam classifiers).

In contrast to one-hot encoding, the embedding produced by word2vec (Lample et al., 2016) represents words as multidimensional continuous floating point vectors,

Table 9.1 Example of word vectors

	Dimension name			
Vector name	*Animal*	*Domesticated*	*Fluffy*	*Pet*
Dog	0.24	−0.03	0.01	−0.05
Cat	0.37	0.48	−0.05	0.21
Rat	−0.02	−0.56	0.31	0.06
Rabbit	−0.14	0.12	0.54	−0.08
Elephant	0.98	0.14	−0.07	0.06

where semantically similar words are mapped to similar vectors within the vector space according to a mathematical definition of similarity (such as cosine similarity as defined below). In simpler terms, a word vector is a row of real-valued numbers where the different numbers reflect different dimensions of the word's meaning and where semantically similar words have nearby vectors. This means, for example, that words such as *screen* and *buttons* should have word vectors that are similar to the vector for *laptop* (because of the similarity of their meanings), whereas the word *strawberry* should be quite dissimilar.

Table 9.1 gives an example of a vector word embedding. As each dimension defines a different meaning. If we represent the first dimension to be the meaning or concept of "animal," then each word's weight on that dimension represents how closely it relates to that concept.

The similarity between two vectors is measured by the cosine of the angle between vectors, using the usual cosine formula. For example, the cosine of the angle between the vectors for *dog* and *cat* in Table 9.1 is

$$\cos(dog, cat) = \frac{dog \cdot cat}{||dog||\,||cat||} = \frac{0.24 \cdot 0.37 - 0.03 \cdot 0.48 - 0.01 \cdot 0.05 - 0.05 \cdot 0.21}{\sqrt{(0.24^2 + 0.03^2 + 0.01^2 + 0.05^2)\,(0.37^2 + 0.48^2 + 0.05^2 + 0.21^2)}}$$

9.3.4.2 Word2vec Architecture

The word2vec encoding algorithms are implemented using an artificial neural network (ANN) structure. ANN represents a class of machine learning algorithms that were originally inspired by the neural structure of the human brain. They have been successfully applied to diverse fields, including pattern classification/recognition; system modeling and identification; signal processing; image processing; control systems; and stock market prediction.

ANNs are composed of interconnected neurons, where each neuron corresponds to a nonlinear function applied to one or more inputs to produce an output, which is then multiplied by weights and sent to one or more other neurons. ANNs are commonly represented as graphs, where the graphs' nodes represent neurons and the edges represent inputs and outputs to/from neurons.

Fig. 9.2 Diagram of a single-layer percept

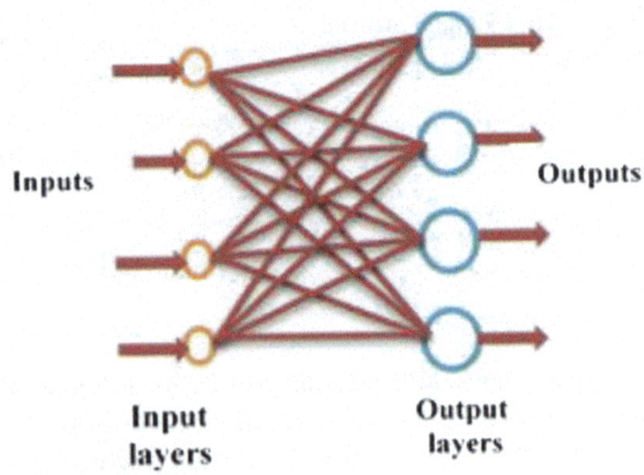

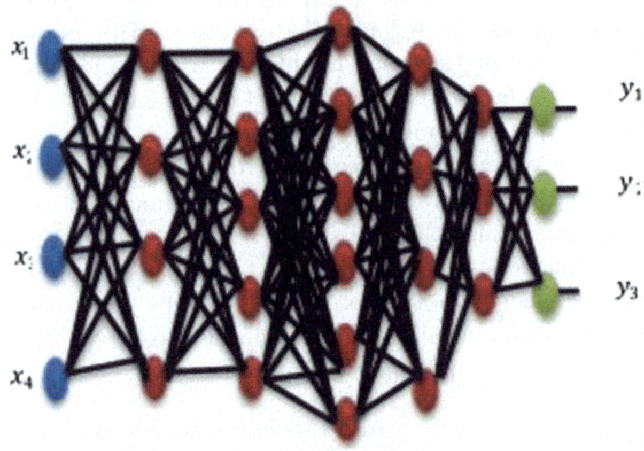

Fig. 9.3 Diagram of a deep neural network (multilayer perceptron)

Figure 9.2 depicts a single-layer perceptron, which is the earliest and most basic type of ANN. The neurons in a single-layer perceptron are classified as either input neurons or output neurons, and the outputs from the input neurons are multiplied by weights and serve as inputs to the output neurons. Single-layer perceptrons are the simplest examples of feed-forward neural networks (FFNN), in which information flows in one direction starting from inputs and terminating in outputs. More complicated FFNN's have one or more hidden layers as shown in Fig. 9.3. These are called multilayer perceptrons (MLP). If the MLP has two or more hidden layers, it is called "deep."

Deep FFNNs can have dozens of layers and are capable of modeling high-level abstractions using multiple layers with more complex structures. They have been applied in many areas of artificial intelligence (AI) such as speech recognition, image recognition, and natural language processing (NLP). However, deep NNs also have significant disadvantages in that they require large amounts of training data, and the training process may be costly in terms of time and computational requirements. In these respects, shallow NNs are preferable to deeper ones, but this advantage may come at the expense of relatively poor performance. The challenge therefore is to find a shallow-network approach that still retains high performance.

Word2Vec uses a shallow NN with one hidden layer to create word embeddings. It includes two different methods for computing embeddings: continuous bag-of-words (CBOW) and skip-gram. These two alternatives correspond to two different methods for training the NN that is used to compute embeddings. In both cases, unsupervised learning can be used to train the NN. This is a strong advantage, because it enables the use of very large, unlabeled datasets for training. The following subsections describe the skip-gram and CBOW embeddings.

9.3.4.3 Continuous Bag-of-Words Model for Obtaining Vector Representation

The CBOW model for creating word embeddings trains the NN by taking context information as input and using it to predict the word within the context. For example, suppose we define the “context” of a word in a sentence as the words immediately before and after the given word. So in the sentence Have a nice day, the context of nice is the set of words {*a, day*}. In the CBOW training scheme, the set {*a, day*} is input to the NN, and the output is compared to the vector representation for nice. In practice, contexts are input as sums of one-hot vectors, as shown in the example in Table 9.2.

The overall CBOW embedding process proceeds as follows. Given a set of texts, a set of (input vector, output vector) pairs are constructed according to the method shown in Table 9.2. The NN is then trained according to conventional gradient descent. At the end of the training process, the input weights to the hidden layer form a N × D matrix, where N is the number of words and D is the number of neurons in the hidden layer. The vector representation of the n’th word is then the n’th row of the matrix.

Table 9.2 Matrix for CBOW example with context width 1. The sentence “This sentence is being converted to input” is converted to seven input vectors for training, using one-hot encoding of the words in each context

	Input vectors									Output
Index	<padding>	This	sentence	is	being	converted	to	inputs	<padding>	vector
1	1	0	1	0	0	0	0	0	0	this
2	0	1	0	1	0	0	0	0	0	sentence
3	0	0	1	0	1	0	0	0	0	is
4	0	0	0	1	0	1	0	0	0	being
5	0	0	0	0	1	0	1	0	0	converted
6	0	0	0	0	0	1	0	1	0	to
7	0	0	0	0	1	0	1	0	1	inputs

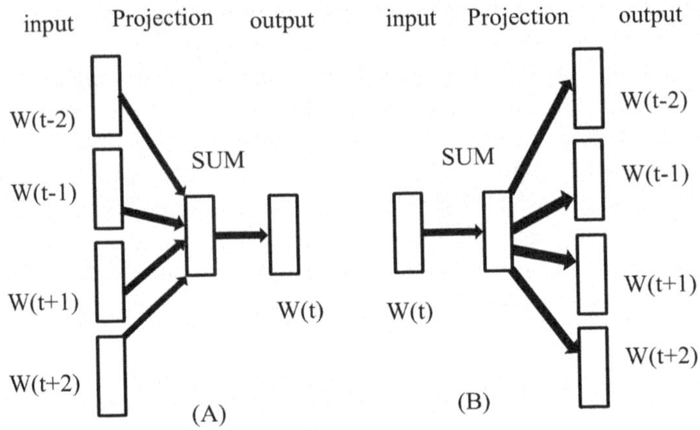

Fig. 9.4 Schematic representations of CBOW (**a**) and skip-gram (**b**) models, in the case where context width is 2 (Mikolov et al., 2013)

9.3.4.4 Skip-Gram Model for Obtaining Vector Representations

Skip-gram differs from CBOW in that instead of using a word's context to predict the word, rather it uses each word in a sentence to predict words in its context. Thus the inputs to the skip-gram NN are one-hot vectors and not combinations. However, the outputs are vector representations of single words, as in CBOW.

The training process for skip-gram is identical to CBOW: input vectors are passed through the NN and compared with the vector representations of the corresponding output vectors. As with CBOW, gradient descent is used to adjust the weights.

9.3.4.5 Comparison Between CBOW and Skip-Gram

Figure 9.4 compares schematic representations of the **CBOW** and skip-gram models, in the case where the context width is 2 (i.e., two words on each side of the given word). The diagram highlights the differences in inputs and outputs: in skip-gram, each base word produces multiple outputs, while in CBOW the information from multiple words is combined to form a single input for each base word. It follows that for a given dataset of texts, skip-gram will produce many more vectors than CBOW (note, e.g., that the number of vectors in Table 9.3 is almost double the number in Table 9.2).

In terms of performance, for frequent words Mikolov (Mikolov et al., 2013) found that for semantic tasks involving frequent words, the CBOW model gave better results, while skip-gram worked better with rare words or in cases where the training dataset is small. For Twitter, previous research seems to favor skip-gram over CBOW (Soliman et al., 2017; Kou et al., 2015).

Table 9.3 Matrix for skip-gram example with context width 1. The sentence "This sentence is being converted to input" is converted to 12 input vectors for training, using one-hot encoding of the words in each context

	Input vectors									Output
Index	<padding>	This	sentence	is	being	converted	to	inputs	<padding>	vector
1	0	1	0	0	0	0	0	0	0	sentence
2	0	0	1	0	0	0	0	0	0	this
3	0	0	1	0	0	0	0	0	0	is
4	0	0	0	1	0	0	0	0	0	sentence
5	0	0	0	1	0	0	0	0	0	being
6	0	0	0	0	1	0	0	0	0	is
7	0	0	0	0	1	0	0	0	0	converted
8	0	0	0	0	0	1	0	0	0	being
9	0	0	0	0	0	1	0	0	0	to
10	0	0	0	0	0	0	1	0	0	converted
11	0	0	0	0	0	0	1	0	0	input
12	0	0	0	0	0	0	0	1	0	to

9.3.5 FastText

This section explains the FastText model. So the more data the models are trained on, the better the results can be. Its performance is almost as good as deep convolution neural network (DCNN) but is orders of magnitude faster in training and evaluation than DCNN. The big difference is that FastText is a shallow network whereas word embeddings to solve the tag prediction task.

Figure 9.5 represents the model architecture of FastText for a sentence with features $x_1, \ldots\ldots, x_N$. The features are embedded and averaged to form the hidden variable. Less than a minute, it is the time that FastText takes to train the dataset. Because it doesn't use pre-trained word embeddings, it has a 1% difference in accuracy (Joulin et al. 2017).

FastText can learn vectors for sub-parts of words, for example, the words "discuss," "discussion," and "discusser" may all have different contexts within a corpus of texts but produce similar vectors in FastText. Output neurons are in bisection with the labels in the supervised mode, but input neurons that have two neurons of the network are the same and that rather than the words in the vocabulary. The weakness of FastText is the representation of un-meaningful words (such as "so" or "the") which contributes equally to the representations of words that are important to the classification.

JFastText is Java wrapper for FastText (Vinhkhuc, 2016), which includes all capabilities of FastText's command-line interface. It also provides the API for loading trained models from files to do label prediction in memory.

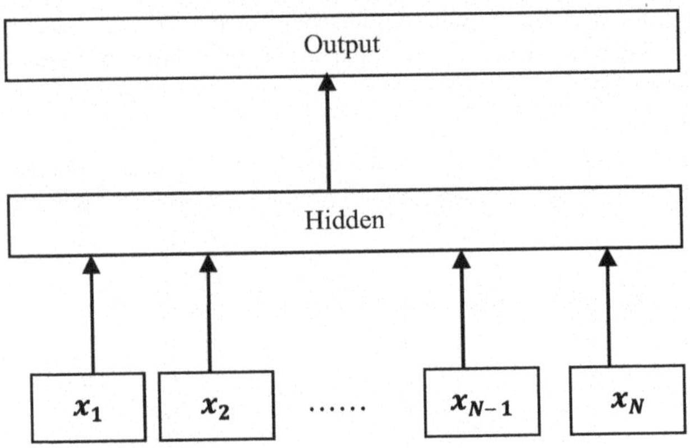

Fig. 9.5 Model architecture of FastText (Mikolov et al., 2013)

9.3.6 Storm

9.3.6.1 Overview of Storm

Storm is an open source distributed real-time data processing system. It was developed by Nathan Marz. Before Storm, Hadoop and Cassandra were commonly used, but they use batch processing which takes all data as input at once, processes it, and then writes a large output. It makes the system slow due to high latency. In contrast, Storm can perform real-time, continuous computations on large amounts of streaming data with high fault tolerance. Storm can support many programming languages, which it makes it more flexible. Its robustness guarantees data processing even if any of the connected nodes in the cluster dies or messages are lost. Storm's strengths are evidenced by the fact that Twitter uses Storm.

9.3.6.2 Components and Performance Metrics of Storm

This section will define the main components of Storm. These components make Storm as an abstraction layer suitable for development

- *Tuples* are the main data structures in Storm. A tuple consists of values passed to a Storm cluster. Tuples compose the data streams processed by Storm.
- *Spouts* are used to emit data streams into a topology. A spout can be created for any stream defined in the configuration. A spout reads data as input from an external source and then passes it to be processed in the application. For example, spouts can connect to the Twitter Streaming API, which is a web service tool providing Twitter to get the tweets submitted by people in real time. There are two types of spouts: reliable and unreliable. If there are any failures in receiving data, a reliable spout will replay the omitted tuples.

- *Bolts* act as logical units of the application, which perform processing tasks on streams and produce output that is delivered to another processing stage. Bolts can perform computations like filtering, streaming aggregation or joining, and updating databases. "IBolt" is the core interface for implementing bolts.

Several metrics are used to measure the performance of a running Storm platform. The performance metrics used in this research are as follows:

- *Complete latency* is defined for spouts as the average amount of time it takes for acknowledgment (ack) or fail to be called for a tuple after it was emitted by the input source. If acking is disabled, this metric is likely to be blank or 0 for all values and should be ignored.
- Capacity is the percent of time that the bolts are idle.
- Process latency is defined for bolts as the average amount of time between the call to start processing a tuple, to notification of ack or fail by the bolt.
- execute-latency is defined for bolts as the average amount of time spent in the call to the execute method. The higher the execute latency, the lower the throughput of tuples per bolt instance.

9.4 Materials and Methods

9.4.1 Overview

We built our FSD model over Storm to analyze the tweets that have local and distributed modes. When tasks in local mode run without any error, we can change the model to be distributed with few changes in settings. In distributed mode, we implemented our FSD with two clusters using two machines running the Ubuntu operating system on a core i5, with 16 GB RAM. One of the machines is virtual.

In this research we compared accuracies obtained using TF-IDF, mTF-IDF, and FastText models. TF-IDF and mTF-IDF were described in Sects. 9.3.2 and 9.3.3. Section 9.4.2 describes our FastText implementation and Sect. 9.4.3.

9.4.2 FastText First Story Detection Implementation

Figure 9.6 shows a system diagram for the FSD based on FastText. As shown in the figure, a word embedding model is a key component of the system. The creation of this model is described in the following subsection.

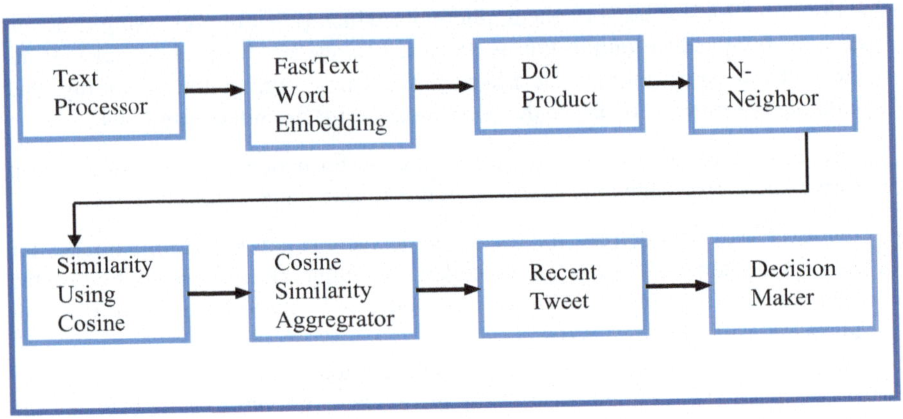

Fig. 9.6 FastText-based system for first story detection

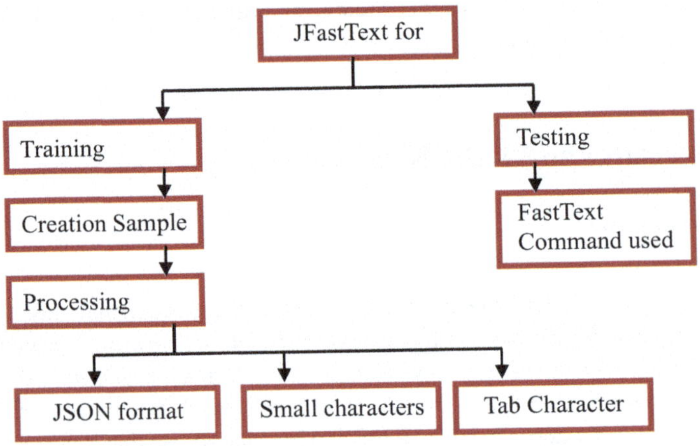

Fig. 9.7 Flow diagram for word embedding using FastText

9.4.2.1 Creating the Word Embedding Using FastText

Figure 9.7 outlines the process used for creating the word embedding in the FastText-based FSD system. The training was implemented using JFastText, which is a Java interface for FastText. The diagram shows both training and testing procedures.

For training and testing 20,000 tweets were retrieved, we part them into two parts: one for training and the other for testing. Training was accomplished using 13,000 tweets, which were extracted and stored in the file `train_file.txt` using the following JFastText command:

```
Head -13000 tweets.txt>train_file.txt
```

The tweets are in *JSON* format. Since FastText uses only the text of tweets, the text was extracted from tweets and stored in the file `json.txt` using the following command:

```
cat train_file.txt | jq -r .text >json.txt
```

Three different FastText models were used, in order to compare performance. To create the first model (`model_a`), FastText was applied directly to `json.txt`, using the following command:

```
./FastText skipgram -input /path/of/json.txt -output model_a -lr
0.025 -dim 10 -ws 5 -epoch 5 -minCount 100 -loss hs -bucket
2000000 -minn 1 -maxn 6 - thread 4 -t 1e-4 -lrUpdateRate 100
```

To create the second FastText model, the tweets in `json.txt` were converted to lower cause and stored in `json_small_char.txt` using the following command:

```
tr '[A-Z]' '[a-z]' <json.txt > json_small_char.txt
```

The above `skipgram` command was then run on `json_small_char.txt` to create `model_b`.

To create the third FastText model, a sequence number was added before each tweet in `json_small_char.txt`, and a Tab character was appended to the end. The resulting tweets were stored in the file `tab.txt` using the following command:

```
awk '{printf("%d\t%s\n",NR,$0);}'<json_small_char.txt>tab.txt
```

The same `skipgram` command was then run on `tab.txt` to create `model_c`.

The three FastText models (henceforth denoted as a, b, c) were tested by applying the model binary files from the three models to the testing data set of 7000 tweets.

We loaded the created trained model files during the initialization within the Storm environment by using the directory path of each model file binary in the constructor of a Bolt class in local mode. As the library does not support serialization in distributed mode, the trained model files can be loaded in the preparation method of Bolt class.

9.4.3 Description of Dataset

We retrieved tweets as a dataset from Twitter4j that has tweets with different languages like English, Arabic, French, etc. We used around 100 MB of data from the Twitter API, which corresponds to about 20,000 tweets. All implementations were run on the same corpus, which was retrieved on July 26, 2017, during President Trump's weekly address.

Tweets were cleaned according to the following commonly used steps:

- Removing web links
- Removing hashtags
- Removing quotes (@moh, etc.)
- Removing punctuations, symbols, and numbers
- Removing stop words, i.e., frequently used words which do not contribute to the meaning and are useless for text classification
- Removing non-English tweets
- Conversion to lowercase

The words of each tweet were converted to vectors using word embeddings: thus each tweet was represented as a set of vectors.

9.5 Results

Table 9.4 shows real sample results obtained after submitting the topology to Storm UI using the original implementation, the first enhanced implementation, and three different versions of the second enhanced implementation, where all the implementations run on a corpus retrieved on July 26, 2017, during President Trump's weekly address. The table shows values and timings of some tweets similarities with the tweets detected as first stories by our modified implementations that use the mTF-IDF, FastText with our tries, and the original implementation that uses the TF-IDF.

As we see from the results, the use of all algorithms except TF-IDF enabled the system to detect tweets in the case of T6 on the same topic with earlier timings, i.e., they are more accurately detected as first stories in our three tries in FastText. Also, in the case of T2, T3, T4, and T5, the tweet identified by TF-IDF and mTF-IDF is not on topic, while FastText with Encoding and without Encoding 1 is more accurate than TF-IDF and mTF-IDF that comes earlier. From the above, we see that the use of FastText with and without is promising as it gives better (or at least equal) accuracy than the traditional TF-IDF and mTF-IDF. Tables 9.5, 9.6, 9.7, 9.8, and 9.9 show the results we got after submitting the topology to the Storm UI for both FSD with TF-IDF, FSD with mTF-IDF, FSD with FastText with Encoding, and FSD with FastText without Encoding 1 and 2. We can see also complete latency for spout, capacity, and process latency for bolts. The "window" column refers to the past period of time for which the statistics apply. We graph the results also in Figs. 9.8, 9.9, 9.10, and 9.11.

Complete latency for spout in TF-IDF (see Table 9.5) is slightly higher than mTF- IDF in Tables 9.7, while complete latency in method b with FastText is higher than others in Tables 9.5, 9.6, 9.7, and 9.8. This is expected as the FastText with Encoding calculation is a bit more complex than others, but the increased time is very small and does not significantly affect completion time. We used Apache-Storm version 0.10.2, which can display the running topology and find data bottlenecks

Table 9.4 Comparison between similarity results for TF-IDF, mTF-IDF, and three FastText algorithms

	Input Tweet	FS detected by (TF-IDF)	FS detected by (mTF-IDF)	FS detected by FastText with a	FS detected by FastText with b	FS detected by FastText with c
T1	1	2	3	4	5	6
Time	Jul 28 23:51 :02	Jul 28 23:33:39	Jul 28 23:33:39	Jul 28 23:50:50	Jul 28 23:50:47	Jul 28 23:50:50
Score	X	0.99573537118 3831	0.99573537118 38,309	1.0	1.0	1.0
In topic	X	Yes	Yes	Yes	Yes	Yes
T2	7	8	9	10	11	12
Time	Jul 28 23:34 :19	Jul 28 23:34:11	Jul 28 23:34:11	Jul 28 23:34:15	Jul 28 23:34:15	Jul 28 23:34:14
Score	X	0.27046818372 726,944	0.27046818372 726,944	1.0	1.0	1.0
In topic	X	No	No	Yes	Yes	Yes
T3	13	14	15	16	17	18
Time	Jul 28 23:47 :26	Jul 28 23:31:53	Jul 28 23:39:58	Jul 28 23:47:24	Jul 28 23:47:24	Jul 28 23:47:21
Score	X	0.16770684329 69,643	0.15186554149 84,823	1.0	1.0	1.0
In topic	X	No	No	Yes	Yes	Yes
T4	19	20	21	22	23	24
Time	Jul 28 23:29 :47	Jul 28 23:29:44	Jul 28 23:29:44	Jul 28 23:29:12	Jul 28 23:29:43	Jul 28 23:28:01

(continued)

Table 9.4 (continued)

	Input Tweet	FS detected by (TF-IDF)	FS detected by (mTF-IDF)	FS detected by FastText with a	FS detected by FastText with b	FS detected by FastText with c
Score	X	0.20380080466 133,663	0.20380080466 133,665	0.999383828 695,681	0.9975207 81,007,847 5	1.0
In topic	X	No	No	Yes	Yes	Yes
T5	25	26	27	28	29	30
Time	Jul 28 23:32 :01	Jul 28 23:29:02	Jul 28 23:29:02	Jul 28 23:29:02	Jul 28 23:29:02	Jul 28 23:29:02
Score	X	0.13220127183 280,617	0.13220127183 280,614	1.0	1.0	1.0
In topic	X	No	No	Yes	Yes	Yes
T6	31	32	33	34	35	36
Time	Jul 2823:52:06	Jul 28 23:28:00	Jul 28 23:34:15	Jul 28 23:34:15	Jul 28 23:34:15	Jul 28 23:34:15
Score	X	0.25283697224 9691	0.15049342539 467,275	1.0	1.0	1.0

Table 9.5 Stats results for topology with TF-IDF

	Windows	Complete Latency	
	10m 0s	667.118	
3h 0m 0s		345.848	
1d 0m 0s		345.848	
All-time		345.848	
Spouts(All Time)			
	Id	Complete	
	spout0	345.848	
Bolts(All Time)			
Id	Capacity	ExecuteLatency	ProcessLatency
b0	0.012	0.041	0.588
b1	0.022	0.802	0.790
b2	0.207	33.907	34.942
b3	0.006	0.986	0.981
b4	0.017	0.099	0.896
b5	0.416	74.277	76.797
b6	0.002	0.039	3.349
b7	0.001	0.117	85.838
b8	0.000	0.081	0.055
b9	0.019	0.084	0.063

Table 9.6 Stats Results for topology with FastText with method a

Window		Complete Latency	
10m 0s		1249.053	
3h 0m 0s		507.017	
1d 0m 0s		507.017	
All-time		507.017	
Spouts(All Time)			
	Id	Complete	
	spout0	507.017	
Bolts(All Time)			
Id	Capacity	ExecuteLatency	ProcessLatency
b0	0.002	0.211	124.462
b1	0.008	0..061	0.201
b2	0.002	0.127	0.093
b3	0.092	0.094	0.420
b4	0.003	1.494	1.440
b5	0.004	0.043	0.019
b6	0.743	174.000	182.130
b7	0.000	0.143	0.097
b8	0.003	0.115	0.089
b9	0.003	0.052	0.124

Table 9.7 Stats Results for topology with mTF-IDF

	Window	Complete Latency	
	10m 0s	626.863	
3h 0m 0s		381.742	
1d 0m 0s		381.742	
All-time		381.742	
Spouts(All Time)			
		Id	
	spout0	381.742	
Bolts(All Time)			
Id	Capacity	ExecuteLatency	ProcessLatency
b0	0.437	85.883	89.110
b1	0.016	0.102	0.954
b2	0.004	1.150	1.147
b3	0.021	0.857	0.845
b4	0.002	0.044	3.759
b5	0.193	35.227	36.864
b6	0.001	0.128	94.790
b7	0.010	0.041	0.576
b8	0.019	0.088	0.066
b9	0.000	0.109	0.083

Table 9.8 Stats Results for topology with FastText with method b

	Window	Complete Latency	
	10m 0s	1075.828	
3h 0m 0s		451.380	
1d 0m 0s		451.380	
All-time		451.380	
Spouts(All Time)			
	Id	Complete	
	spout0	451.380	
Bolts(All Time)			
Id	Capacity	ExecuteLatency	ProcessLatency
b0	0.003	0.101	0.077
b1	0.001	0.120	0.082
b2	0.753	155.005	161.149
b3	0.003	1.211	1.161
b4	0.165	0.119	0.353
b5	0.002	0.045	0.114
b6	0.003	0.038	0.016
b7	0.001	0.176	0.144
b8	0.006	0.053	0.184
b9	0.001	0.209	110.642

Table 9.9 Stats Results for topology with FastText with method c

	Window	Complete Latency	
	10m 0s	1119.919	
3h 0m 0s		465.720	
1d 0m 0s		465.720	
All-time		465.720	
Spouts(All Time)			
	Id	Complete	
	spout0	465.720	
Bolts(All Time)			
Id	Capacity	ExecuteLatency	ProcessLatency
b0	0.004	0.040	0.018
b1	0.002	0.103	0.078
b2	0.675	160.460	167.283
b3	0.000	0.111	0.077
b4	0.001	0.205	114.284
b5	0.001	1.123	1.094
b6	0.001	0.045	0.112
b7	0.000	0.392	0.324
b8	0.006	0.051	0.187
b9	0.073	0.085	0.372

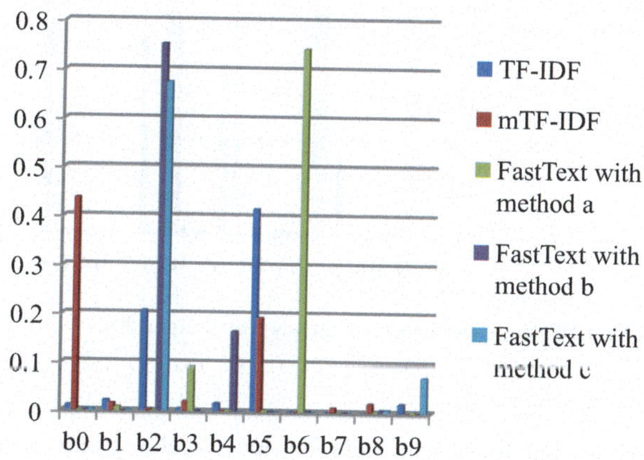

Fig. 9.8 Bolts and Capacity for all algorithms

in the running Storm application by clicking the *Show Visualization* button in the Storm UI. Visual representations of the FSD topology in original, modified, FastText with Encoding, and FastText without Encoding1 and 2 implementations are shown in Figs. 9.12, 9.13, 9.14, 9.15, and 9.16, respectively. *Thicker lines* between components denote larger data flows. A blue component represents the first component in the topology, such as the spout in Figs. 9.12, 9.13, 9.14, 9.15, and 9.16. Topology components have colors that indicate capacity status: red components

Fig. 9.9 Bolts and Execute Latency for all algorithms

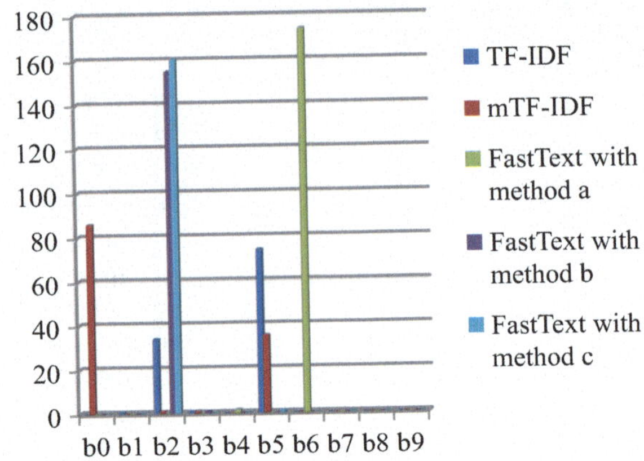

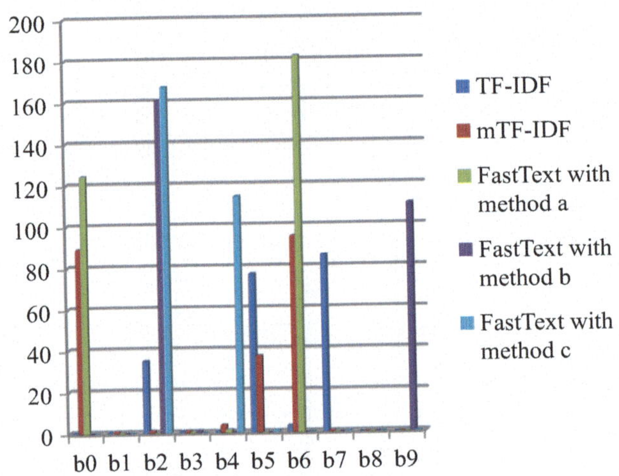

Fig. 9.10 Bolts and Process Latency for all algorithms

mean that there is a data bottleneck and green components denote components operating within capacity. Milliseconds time denotes the time of each bolt.2 A comparison between all algorithms in terms of capacity, execute latency, and process latency is given in Figs. 9.17, 9.18, and 9.19, respectively. We numbered the tweets to make Table 9.4 more readable in the Appendix.

Tables 9.4, 9.5, 9.6, 9.7, and 9.8 show the difference between Bolts and Capacity, Execute Latency, and Process Latency, respectively, for (TF-IDF), (mTF-IDF), (FastText with method a), and (FastText with method b and c).

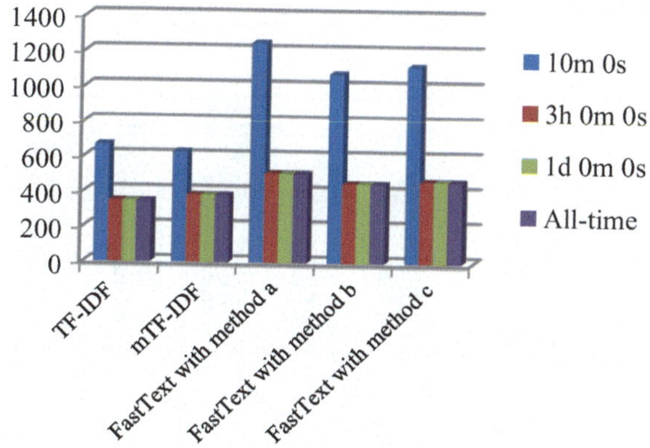

Fig. 9.11 Topology state and Complete Latency for all algorithms

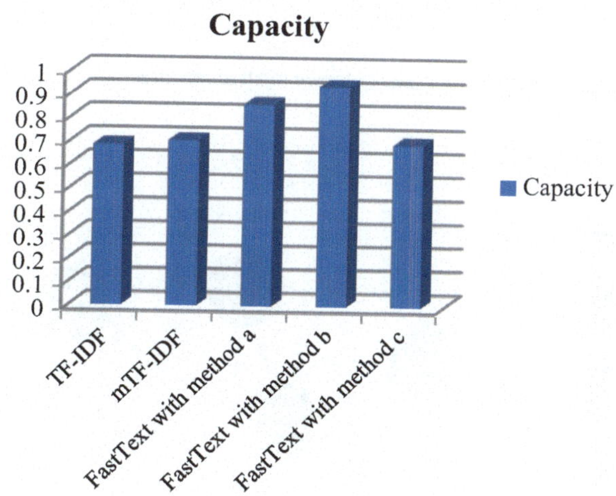

Fig. 9.12 Capacity between all algorithms

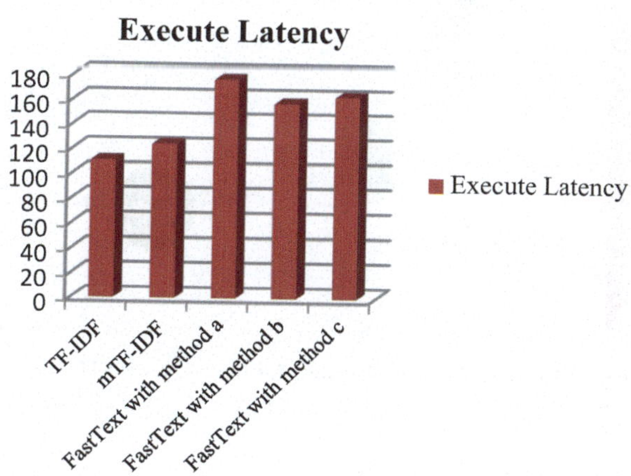

Fig. 9.13 Execute Latency between all algorithms

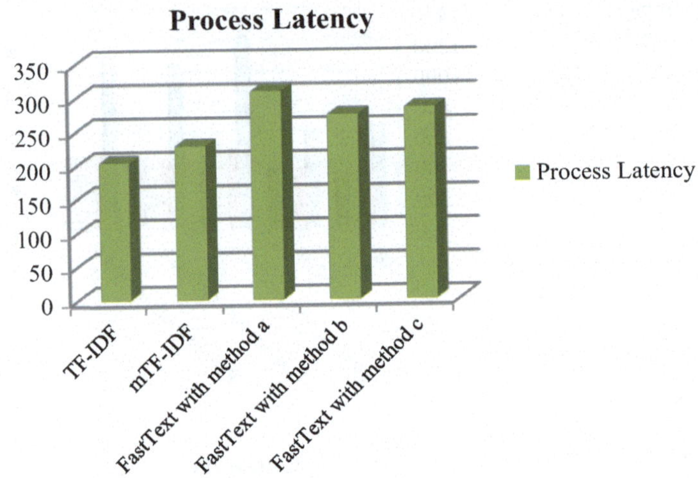

Fig. 9.14 Process Latency between all algorithms

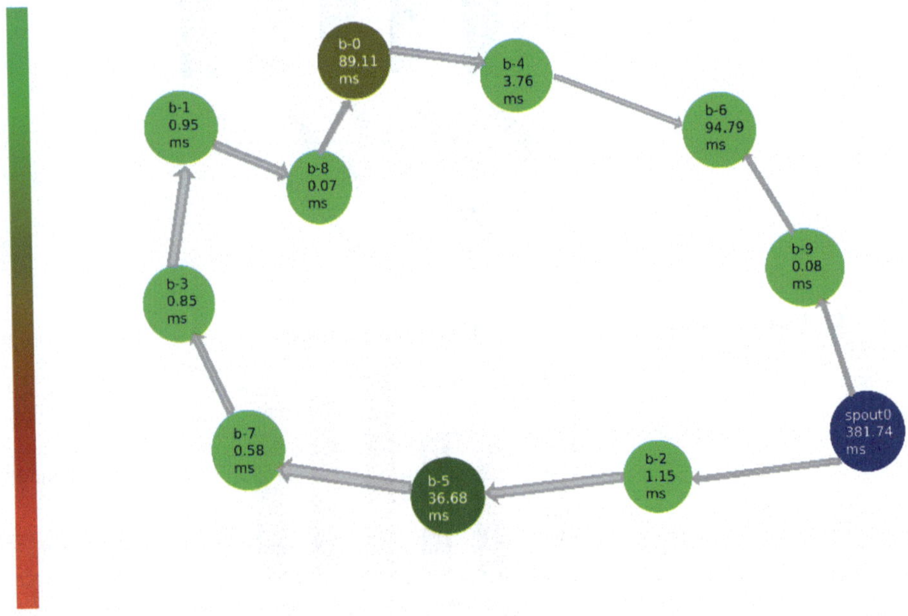

Fig. 9.15 Visualization of results for topology with TF-IDF

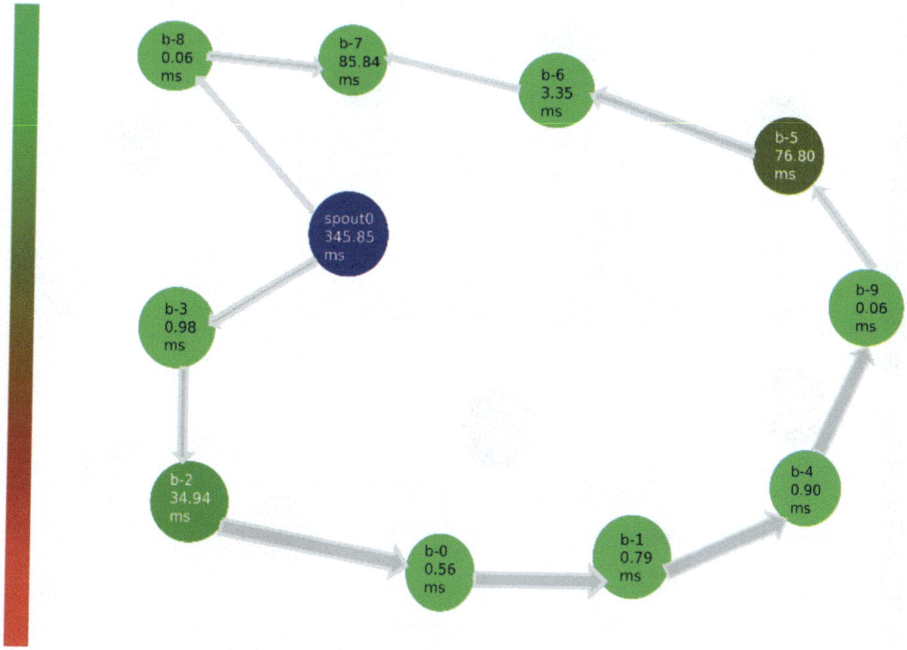

Fig. 9.16 Visualization of results for topology with mTF-IDF

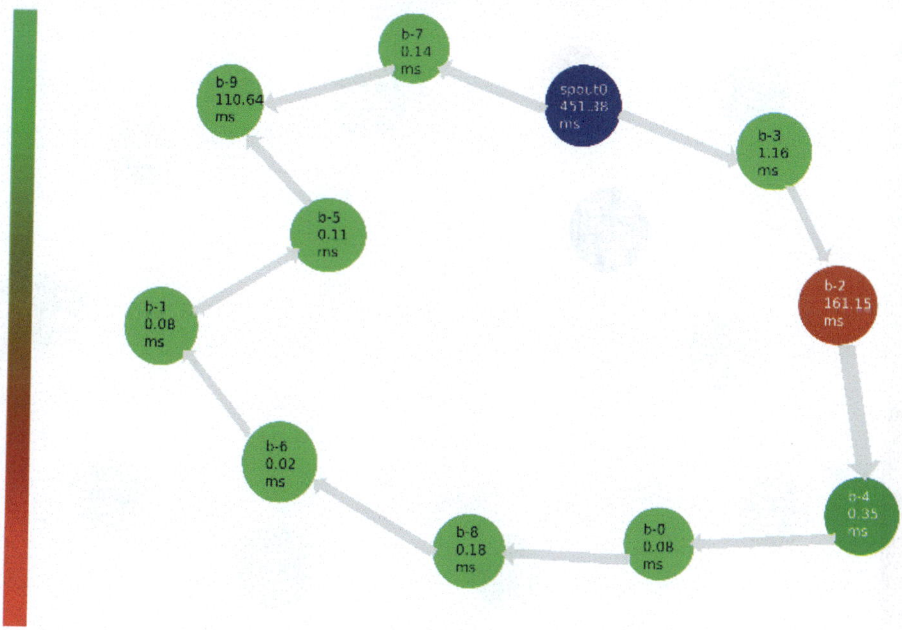

Fig. 9.17 Visualize results for topology with FastText with method a

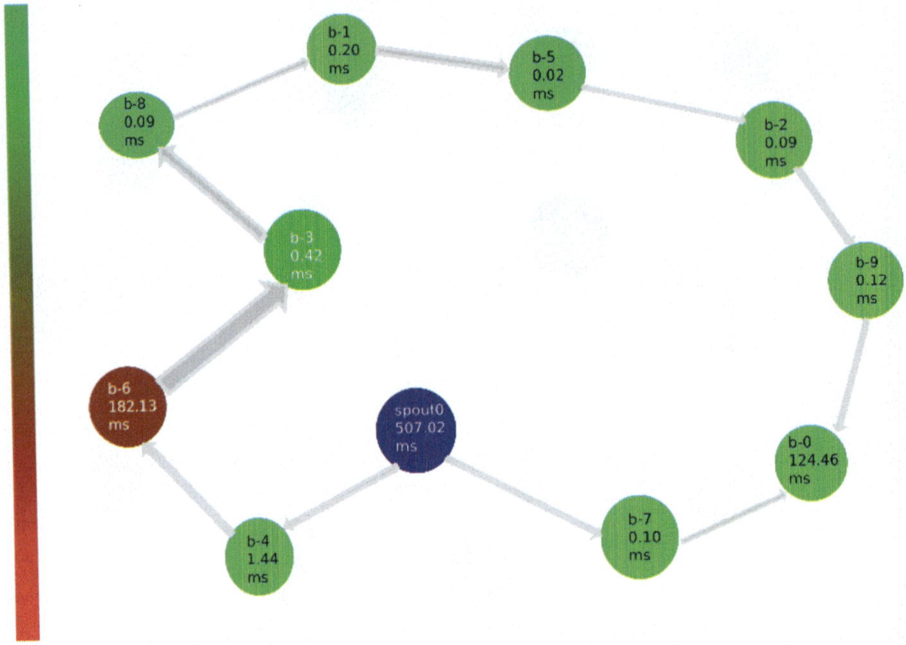

Fig. 9.18 Visualization of results for topology with FastText with method b

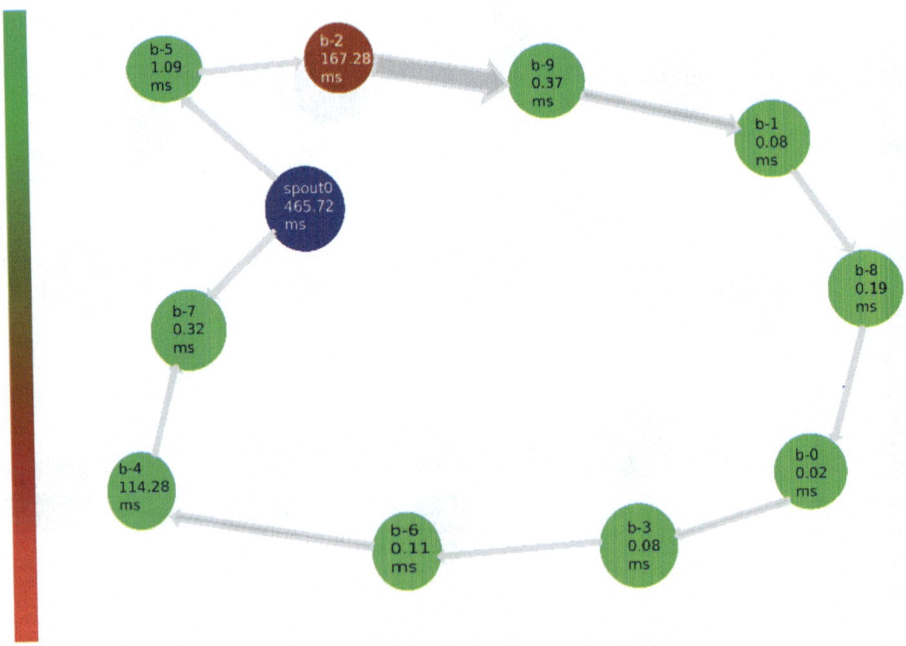

Fig. 9.19 Visualization of results for topology with FastText with method c

The following results in Table 9.10 for testing the 6000 tweets with the same configurations in submitting the topology in Storm. Tables 9.11, 9.12, and 9.13 show the Complete, Execute, and Process Latency for three topologies in testing mode with its graph in Figs. 9.20, 9.21, 9.22, and 9.23 with other algorithms TF-IDF and mTF-IDF. We also figured a comparison between all algorithms with its Capacity, Execute Latency, and Process Latency in Figs. 9.24, 9.25, and 9.26, respectively. We visualized the representation of the FSD topology in original, modified, FastText in Figs. 9.27, 9.28, and 9.29, respectively. We numbered the tweets to make Table 9.10 more readable in the Appendix.

9.6 Conclusion and Future Work

Distributed real-time platforms have advantages such as a high degree of parallelism and guaranteed real-time execution, both required by FSD systems. Storm is a distributed real-time platform that is an open-source used for stream processing. It is flexible and scalable and delivers high performance. Some systems of FSD that are built using Storm used traditional TF-IDF to measure the similarity between tweets. For enhancing accuracy, we used the char2vec or FastText model which is more efficient. Our results show that the result of FastText was better than mTF-IDF and TF-IDF in time and accuracy.

In the future, we can improve our system for enhancing accuracy by using other algorithms for subtext similarities. We intend to build a set of parallel supervised detectors followed by a final decision maker to enhance the overall performance, as we used the unsupervised clustering method in this chapter. Also, we intend to use subword-based distance measurement methods instead of word based methods. After that, we plan to build a customized version of the system proposed here for Arabic tweets, in which the specific features of the Arabic language are embedded in pre and post-processing.

Table 9.10 Comparison between Similarity result for TF-IDF, mTF-IDF, and FastText Algorithm

	Input Tweet	FS Detected by (TF-IDF)		FS Detected by (mTF-IDF)		FS Detected by FastText (method a)		FS Detected by FastText (method b)		FS Detected by FastText (method c)	
T1	37	38		39		40		41		42	
Time	Jul 2823:51:02 23:5 1:02	Jul 23:33:39	28	Jul 23:33:39	28	Jul 23:28:44	28	Jul 23:28:44	28	Jul 23:28:44	28
Score	X	0.9957353711 83,831		0.9957353711 838,309		1.0		1.0		1.0	
		83,831		838,309							
In Topic	X	Yes		Yes		Yes		Yes		Yes	
T2	43	44		45		46		47		48	
Time	Jul 28 4:19 23:3	Jul 23:34:11	28	Jul 23:34:11	28	Jul 23:34:15	28	Jul 23:34:15	28	Jul 23:34:15	28
Score	X	0.270468183 72,726,944		0.270468183 72,726,944		1.0		1.0		1.0	
In Topic	X	No		No		Yes		Yes		Yes	
T3	49	50		51		52		53		54	
Time	Jul 2823:47:26 23:4 7:26	Jul 23:31:53	28	Jul 23:39:58	28	Jul 23:47:24	28	Jul 23:47:24	28	Jul 23:47:21	28

Table 9.10 (continued)

Score	X	0.167706843 2,969,643 2,969,643		0.151865541 4,984,823 4,984,823		1.0		1.0		1.0	
In Topic	X	No		No		Yes		Yes		Yes	
T4	55	56		57		58		59		60	
Time	Jul 2823:29:47	Jul 23:29:44	28	Jul 23:29:44	28	Jul 23:29:12	28	Jul 23:29:42	28	Jul 23:29:25	28
Score	X	0.203800804 66,133,663		0.203800804 66,133,665		0.999013988 9,208,077		0.999050141 5,862,676		0.999020977 6,353,844	
In Topic	X	No		No		Yes		Yes		Yes	
T5	61	62		63		64		65		66	
Time	Jul 2823:32:01	Jul 23:29:02	28	Jul 23:29:02	28	Jul 23:31:55	28	Jul 23:31:55	28	Jul 23:31:55	28
Score	X	0.132201271 83,280,617		0.132201271 83,280,614		1.0		1.0		1.0	
In Topic	X	No		No		Yes		Yes		Yes	
T6	67	68		69		70		71		72	
Time	Jul 2823:52:06	Jul 23:28:00	28	Jul 23:34:15	28	Jul 23:34:15	28	Jul 23:34:15	28	Jul 23:34:15	28
Score	X	0.252836972 249,691		0.150493425 39,467,275		1.0		1.0		1.0	
InTopic	X	No		No		Yes		Yes		Yes	

Table 9.11 Stats Results for topology with FastText with method a

	Window	Complete Latency	
	10m 0s	1138.750	
3h 0m 0s		458.837	
1d 0m 0s		458.837	
All-time		458.837	
Spouts(All Time)			
	Id	Complete	
	spout0	458.837	
Bolts(All Time)			
Id	Capacity	ExecuteLatency	ProcessLatency
b0	0.003	1.470	1.413
b1	0.001	0.130	0.088
b2	0.782	156.328	162.616
b3	0.005	0.043	0.018
b4	0.164	0.113	0.361
b5	0.002	0.249	112.015
b6	0.001	0.052	0.124
b7	0.003	0.144	0.118
b8	0.008	0.056	0.185
b9	0.001	0.184	0.133

Table 9.12 Stats Results for topology with FastText with method b

	Window	Complete Latency	
	10m 0s	1111.870	
3h 0m 0s		465.297	
1d 0m 0s		465.297	
All-time		465.297	
Spouts(All Time)			
	Id	Complete	
	spout0	465.297	
Bolts(All Time)			
Id	Capacity	ExecuteLatency	ProcessLatency
b0	0.096	84	68
b1	0.007	54	76
b2	0.805	1.627	8.447
b3	0.002	01	0.139
b4	0.002	34	01
b5	0.003	45	19
b6	0.004	42	17
b7	0.000	39	05
b8	0.001	50	19
b9	0.001	28	88

Table 9.13 Stats Results or topology with FastText with method c

	Window	Complete Latency	
	10m 0s	1128.161	
3h 0m 0s		455.479	
1d 0m 0s		455.479	
All-time		455.479	
Spouts(All Time)			
	Id	Complete	
	spout0	455.479	
Bolts(All Time)			
Id	Capacity	ExecuteLatency	ProcessLatency
b0	0.002	0.038	0.017
b1	0.000	0.102	0.070
b2	0.001	1.142	1.136
b3	0.497	157.759	164.364
b4	0.001	0.043	0.106
b5	0.064	0.081	0.360
b6	0.001	0.181	111.668
b7	0.001	0.136	0.113
b8	0.003	0.051	0.169
b9	0.000	0.159	0.107

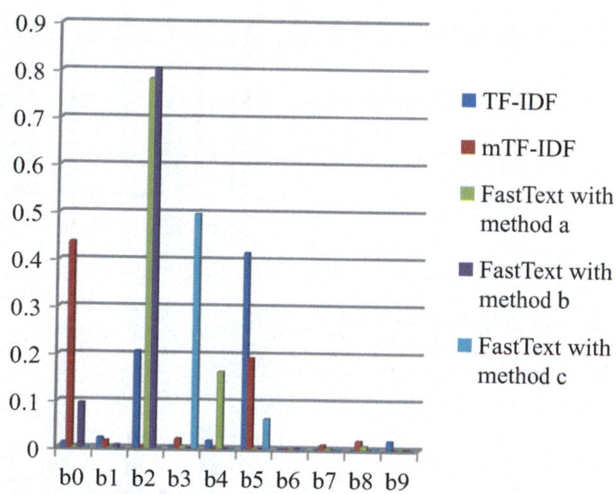

Fig. 9.20 Bolts and Capacity Latency for all algorithm

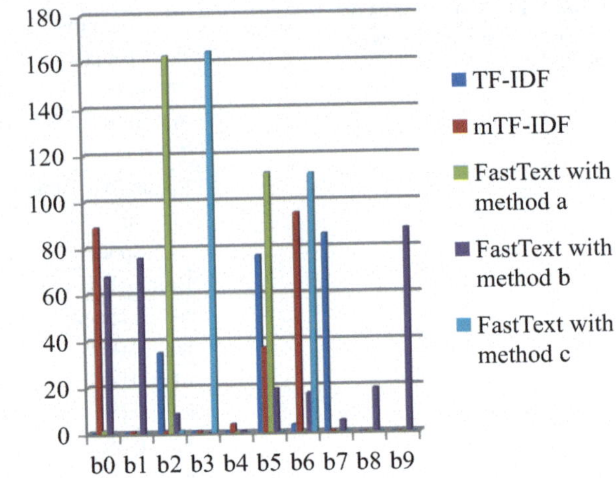

Fig. 9.21 Bolts and Execute Latency for all for all algorithm

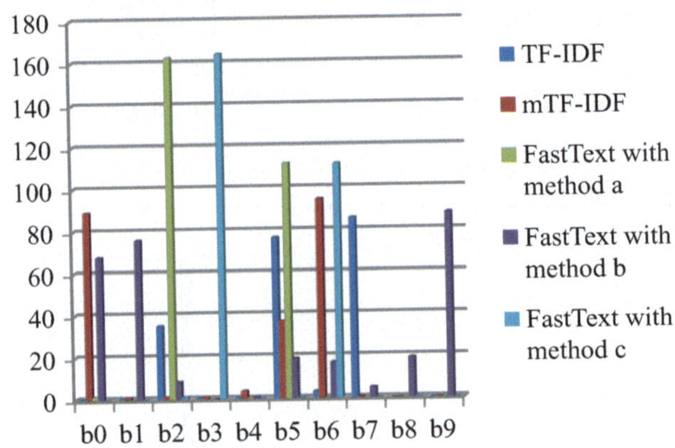

Fig. 9.22 Bolts and Process Latency for all algorithms

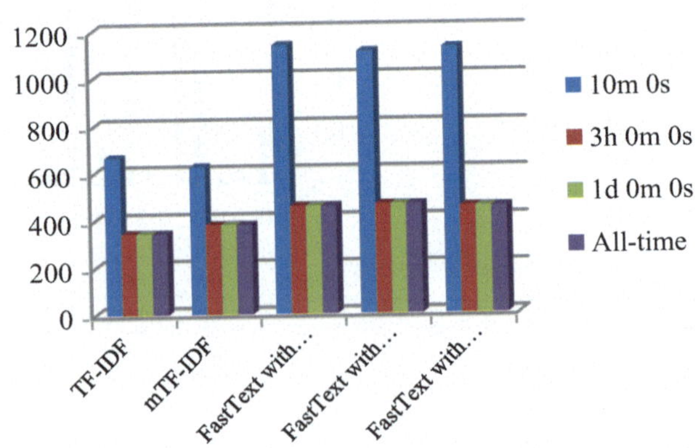

Fig. 9.23 Topology state and Complete Latency for all algorithms

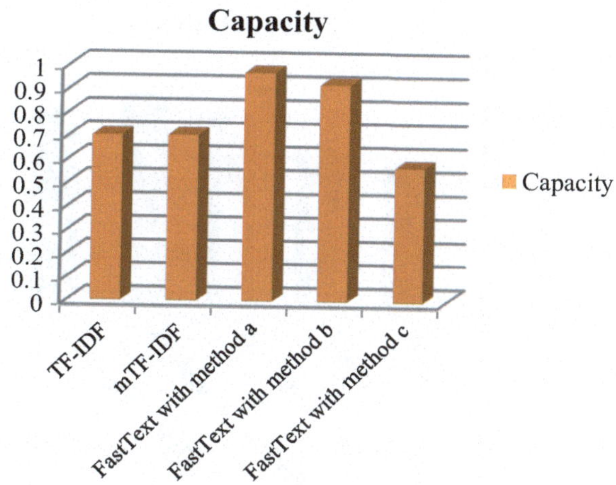

Fig. 9.24 Capacity between all algorithms

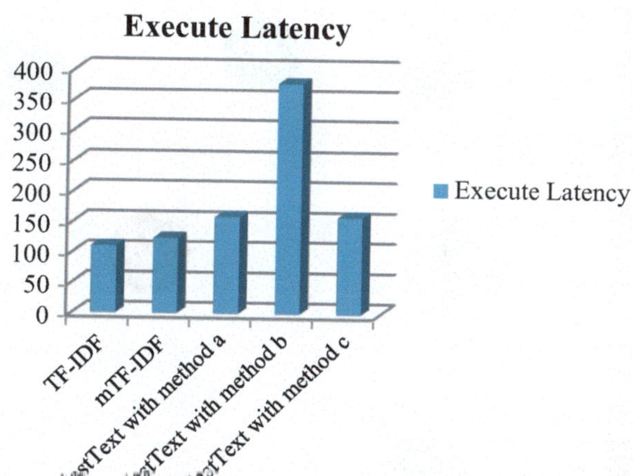

Fig. 9.25 Execute Latency between all algorithms

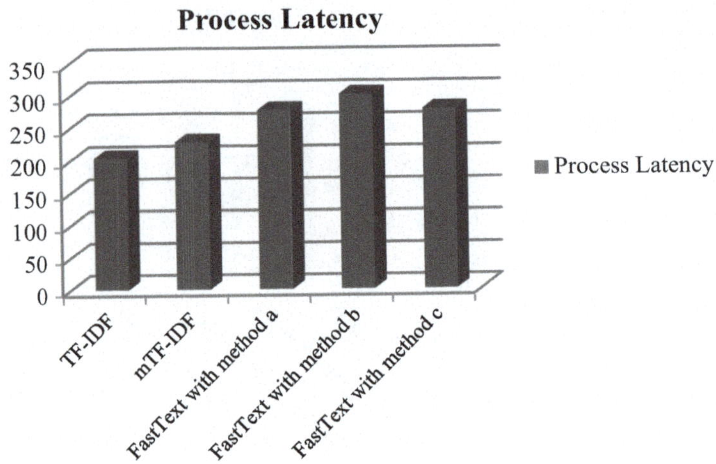

Fig. 9.26 Process Latency between all algorithms

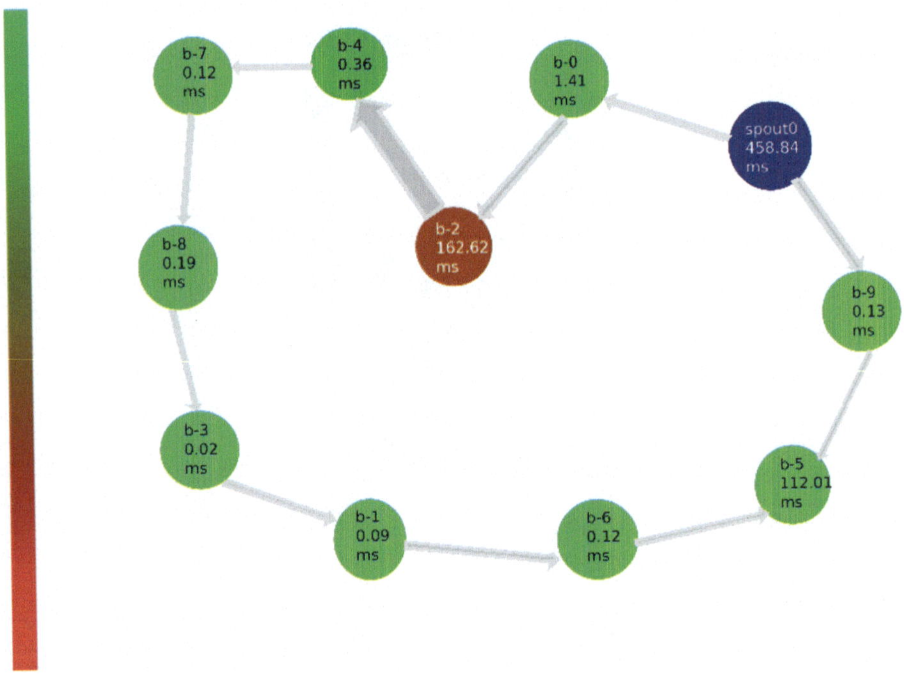

Fig. 9.27 Visualize results for topology with FastText with method a

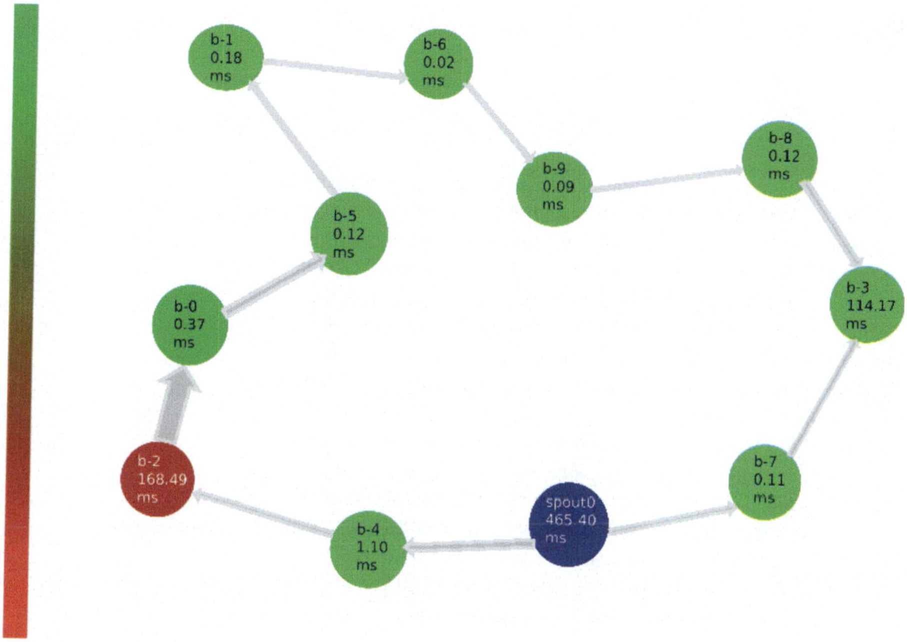

Fig. 9.28 Visualize results for topology with FastText with method b

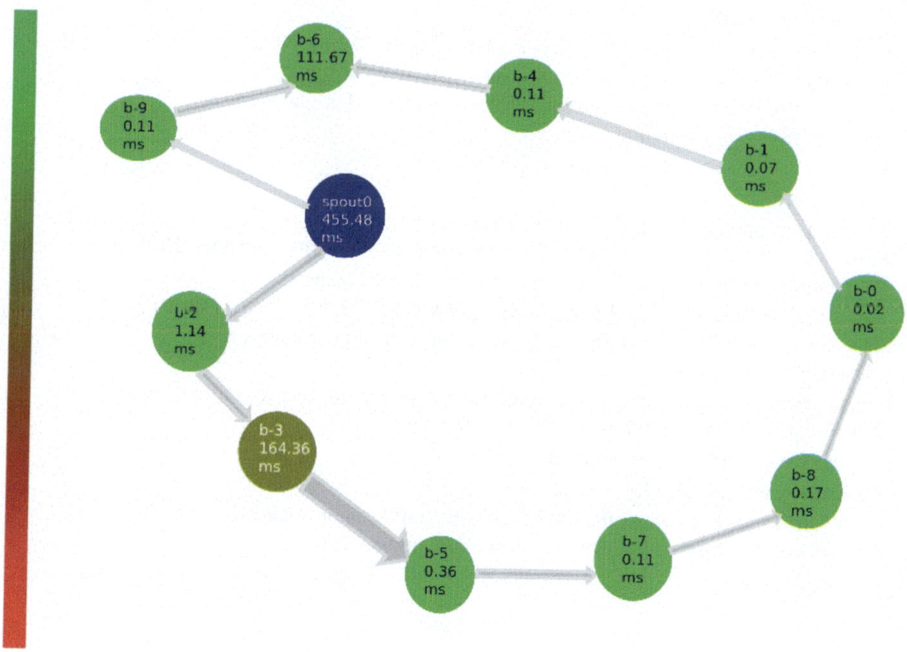

Fig. 9.29 Visualize results for topology with FastText with method c

Appendix

Tweet	No.
jk_rowling: Truly, whom amongst us can forget Trump ordering the @ killing of bin Laden? Or Obama bragging about barging in on na\u2026	1
jk_rowling: Truly, whom amongst us can forget Trump ordering the @ killing of bin Laden? Or Obama bragging about barging in on na\u2026	2
jk_rowling: Truly, whom amongst us can forget Trump ordering the @ killing of bin Laden? Or Obama bragging about barging in on na\u2026	3
jk_rowling: Truly, whom amongst us can forget Trump ordering the @ killing of bin Laden? Or Obama bragging about barging in on na\u2026	4
jk_rowling: Truly, whom amongst us can forget Trump ordering the @ killing of bin Laden? Or Obama bragging about barging in on na\u2026	5
jk_rowling: Truly, whom amongst us can forget Trump ordering the @ killing of bin Laden? Or Obama bragging about barging in on na\u2026	6
RT @mitchellvii: Media: "Will you replace ReincePriebus?" Trump: "Maybe, maybe not." Media: "Can you be specific?" Trump: "I'd rather be G...	7
RT where i'd rather be https:\/\/t.co\/NxaKURHYWu	8
RT where i'd rather be https:\/\/t.co\/NxaKURHYWu	9
RT Trump\u2019s speech in Long Island today was chilling even by his standards. Full statement:\u2026	10
RT Trump\u2019s speech in Long Island today was chilling even by his standards. Full statement:\u2026	11
RT Former Republican Rep. Mike Rogers discussing the Trump White House on CNN: \"It's like watching an octopus put its socks on\:\u2026	12
Trump: Kelly Will Do 'a Fantastic Job,' Priebus 'a Good Man': https://t.co/FBgAewN9z2 via @YouTube	13
RT Good bye, Reince. May you rest in Priebus.s	14
have a good weekend	15
RT Trump Supporters React on Twitter to Trump FIRING Reince! https:\/\/t.co\/DMKdBSwJL3 John F Kelly Of Homeland Security Took Ov\u2026	16
RT Trump Supporters React on Twitter to Trump FIRING Reince! https:\/\/t.co\/DMKdBSwJL3 John F Kelly Of Homeland Security Took Ov\u2026	17
RT GOP Suicidal-Controls Both Houses, 0 Result- Get Rid of Leaders & RINOs https:\/\/t.co\/s41yFip62m #MAGA #Trump\u2026	18
Thank you @SenatorCollins for your leadership & protecting care for millions in ME & across the US! #IStandWithPP https://t.co/J8iC8g94Fh	19
We thank you in advance for your patience!	20
We thank you in advance for your patience!	21
RT #CharlieKirk Talks about the $$MILLIONS GOP Senators raised on the backs of the promise of repealing ObamaCare https\u2026	22

(continued)

Tweet	No.
RT Se cumple un añodesdeque le encontraron 4.6 millones de dólares a Florencia Kirchner. No laburónunca. Lo dejo a tucr\u2026	23
RT Thank you to the folks in the disability communities, the folks in Planned Parenthood, and the millions more who st\u2026	24
RT @grantstern: Donald Trump Jr. blocked me for finding his #TrumpRussia money laundering ties, so I found his Mexican Cartel ties! https:/...	25
RT like my nigga lowkey ... all about his money and all about me can't have it no other way.	26
RT like my nigga lowkey ... all about his money and all about me can't have it no other way.	27
RT Trump openly encourages needless police brutality. This perpetuates the broken trust that makes violence toward and\u2026	28
RT Trump openly encourages needless police brutality. This perpetuates the broken trust that makes violence toward and\u2026	29
RT Trump openly encourages needless police brutality. This perpetuates the broken trust that makes violence toward and\u2026	30
RT @michaelwild2198: @realDonaldTrump @BarackObamaHow about 3 Chief of staff in 6 months Donnie?? Not to mention Flynn, Comey, etc. et\u2026	31
RT WH Changes: \u2022Chief of Staff \u2022Deputy Chief of Staff \u2022National Security Advisor \u2022Press Secretary \u2022Comm Director https:\/\/\u2026	32
RT 1. National Security Advisor 2. FBI Director 3. Press Secretary 4. Chief of Staff 5. Private Lead Counsel All fired\/re\u2026	33
RT 1. National Security Advisor 2. FBI Director 3. Press Secretary 4. Chief of Staff 5. Private Lead Counsel All fired\/re\u2026	34
RT 1. National Security Advisor 2. FBI Director 3. Press Secretary 4. Chief of Staff 5. Private Lead Counsel All fired\/re\u2026	35
RT 1. National Security Advisor 2. FBI Director 3. Press Secretary 4. Chief of Staff 5. Private Lead Counsel All fired\/re\u2026	36
@jk_rowling: Truly, whom amongst us can forget Trump ordering the killing of bin Laden? Or Obama bragging about barging in on na\u2026	37
RT Truly, whom amongst us can forget Trump ordering the killing of bin Laden? Or Obama bragging about barging in on na\u2026	38
RT Truly, whom amongst us can forget Trump ordering the killing of bin Laden? Or Obama bragging about barging in on na\u2026	39
RT So is on a tirade about how Trump ignored a kid in a wheelchair, but she conveniently left out the clip\u2026	40
RT So is on a tirade about how Trump ignored a kid in a wheelchair, but she conveniently left out the clip\u2026	41
RT So is on a tirade about how Trump ignored a kid in a wheelchair, but she conveniently left out the clip\u2026	42
RT @mitchellvii: Media: "Will you replace ReincePriebus?" Trump: "Maybe, maybe not." Media: "Can you be specific?" Trump: "I'd rather be G...	43
RT where i'drather be https:\/\/t.co\/NxaKURHYWu	44
RT where i'd rather be https:\/\/t.co\/NxaKURHYWu	45

(continued)

Tweet	No.
RT Trump\u2019s speech in Long Island today was chilling even by his standards. Full statement:\u2026	46
RT Trump\u2019s speech in Long Island today was chilling even by his standards. Full statement:\u2026	47
RT Trump\u2019s speech in Long Island today was chilling even by his standards. Full statement:\u2026	48
Trump: Kelly Will Do 'a Fantastic Job,' Priebus 'a Good Man': https://t.co/FBgAewN9z2 via @YouTube	49
RT Good bye, Reince. May you rest in Priebus.s	50
have a good weekend	51
RT Trump Supporters React on Twitter to Trump FIRING Reince! https:\/\/t.co\/DMKdBSwJL3 John F Kelly Of Homeland Security Took Ov\u2026	52
RT Trump Supporters React on Twitter to Trump FIRING Reince! https:\/\/t.co\/DMKdBSwJL3 John F Kelly Of Homeland Security Took Ov\u2026	53
RT GOP Suicidal-Controls Both Houses, 0 Result- Get Rid of Leaders & RINOs https:\/\/t.co\/s41yFip62m #MAGA #Trump\u2026	54
Thank you @SenatorCollins for your leadership & protecting care for millions in ME & across the US! #IStandWithPP https://t.co/J8iC8g94Fh	55
We thank you in advance for your patience!	56
We thank you in advance for your patience!	57
the cardinals are overrated	58
RT. The voters elected someone they wanted to disrupt Washington, DC. They did not want business as usual https\u2026	59
Under pressure album hot as shit	60
RT @grantstern: Donald Trump Jr. blocked me for finding his #TrumpRussia money laundering ties, so I found his Mexican Cartel ties! https:/ . . .	61
RT like my nigga lowkey ... all about his money and all about me can't have it no other way.	62
RT like my nigga lowkey ... all about his money and all about me can't have it no other way.	63
If u watch the video, u can see officers behind Trump in full uniform applauding. No excuse. L\u2026 https:\/\/t.co\/jADi5HUFfl	64
If u watch the video, u can see officers behind Trump in full uniform applauding. No excuse. L\u2026 https:\/\/t.co\/jADi5HUFfl	65
If u watch the video, u can see officers behind Trump in full uniform applauding. No excuse. L\u2026 https:\/\/t.co\/jADi5HUFfl	66
RT @michaelwild2198: @realDonaldTrump @BarackObama How about 3 Chief of staff in 6 months Donnie?? Not to mention Flynn, Comey, etc. et\u2026	67
RT WH Changes: \u2022Chief of Staff \u2022Deputy Chief of Staff \u2022National Security Advisor \u2022Press Secretary \u2022Comm Director https:\/\/\u2026	68
RT 1. National Security Advisor 2. FBI Director 3. Press Secretary 4. Chief of Staff 5. Private Lead Counsel All fired\/re\u2026	69
RT 1. National Security Advisor 2. FBI Director 3. Press Secretary 4. Chief of Staff 5. Private Lead Counsel All fired\/re\u2026	70
RT 1. National Security Advisor 2. FBI Director 3. Press Secretary 4. Chief of Staff 5. Private Lead Counsel All fired\/re\u2026	71
RT 1. National Security Advisor 2. FBI Director 3. Press Secretary 4. Chief of Staff 5. Private Lead Counsel All fired\/re\u2026	72

References

Allan, J., Lavrenko, V., & Jin, H. (2000). First story detection in TDT is hard. In *Proceedings of the ninth international conference on Information and knowledge management* (pp. 374–381). ACM.

Allan, J. (2002). *Topic detection and tracking: Event-based information organization*. Kluwer Academic Publishers.

Ballesteros, M., Dyer, C., & Smith, N. A. (2015). Improved transition-based parsing by modeling characters instead of words with LSTMs. In *Conference on Empirical Methods in Natural Language Processing - EMNLP 2015* (pp. 349–359). https://doi.org/10.18653/v1/d15-1041

Benhardus, J., & Kalita, J. (2013). Streaming trend detection in twitter. *International Journal of Web Based Communities, 9*(1), 122–139.

Bojanowski, P., Grave, E., Joulin, A., & Mikolov, T. (2017). Enriching word vectors with subword information. *Transactions of the Association for Computational Linguistics, 5*, 135–146. https://doi.org/10.1162/tacl_a_00051

Brigadir, I., Greene, D., & Cunningham, P. (2014). Adaptive representations for tracking breaking news on twitter. arXiv preprint arXiv:1403.2923.

Cao, K., & Rei, M. (2016). A joint model for word embedding and word morphology. arXiv preprint arXiv:1606.02601.

Elbedwehy, S., Alrahmawy, M., & Hamza, T. (2017). Real time first story detection in twitter using a modified Tf-Idf algorithm. *International Journal of Intelligent Computing and Information Sciences, 17*, 11–31. https://doi.org/10.21608/ijicis.2017.8245

Godin, F., Vandersmissen, B., De Neve, W., & Van de Walle, R. (2015). Multimedia lab@ ACL W-NUT NER shared task: Named entity recognition for Twitter microposts using distributed word representations. ACL-IJCNLP 2015.

https://research.fb.com/blog/2016/08/FastText/

Huddar, M. G., Ramannavar, M. M., & Sidnal, N. S. (2014). Scalable distributed first story detection using storm for twitter data. In *Advances in engineering and technology research (ICAETR), 2014 international conference on* (pp. 1–5). IEEE.

Indyk, P., & Motwani, R. (1998). Approximate nearest neighbors: Towards removing the curse of dimensionality. In *Proceedings of the thirtieth annual ACM symposium on Theory of computing*.

Joulin, A., Grave, E., Bojanowski, P., & Mikolov, T. (2017). Bag of tricks for efficient text classification. In *Proceedings of the 15th Conference of the European Chapter of the Association for Computational Linguistics EACL 2017* (Vol. 2, pp. 427–431). https://doi.org/10.18653/v1/e17-2068

Kali, M., Rousseau, F., Ntoulas, A., & Vazirgiannis, M. (2013). Efficient online novelty detection in news streams. In *International conference on web information systems engineering* (pp. 57–71). Springer Berlin Heidelberg.

Kim, Y., Jernite, Y., Sontag, D., & Rush, A. M. (2016). Character-aware neural language models. In *30th AAAI Conference on Artificial Intelligence. AAAI 2016* (pp. 2741–2749).

Kou, W., Li, F., & Baldwin, T. (2015). Automatic labelling of topic models using word vectors and letter trigram vectors. *Lecture Notes in Computer Science (including Subser. Lect. Notes Artif. Intell. Lect. Notes Bioinformatics), 9460*, 253–264. https://doi.org/10.1007/978-3-319-28940-3_20

Lample, G., Ballesteros, M., Subramanian, S., Kawakami, K., & Dyer, C. (2016). Neural architectures for named entity recognition. In *Proceedings of the 2016 Conference of the North American Chapter of the Association for Computational Linguistics: Human Language Technologies NAACL HLT 2016* (pp. 260–270). https://doi.org/10.18653/v1/n16-1030

McCreadie, R., Macdonald, C., Ounis, I., Osborne, M., & Petrovic, S. (2013). Scalable distributed event detection for Twitter. In *IEEE International Conference on Big Data, Big Data 2013* (pp. 543–549). https://doi.org/10.1109/BigData.2013.6691620

Meetei, L. S., et al. (2019). Automatic extraction of locations from news articles using domain knowledge. In *International conference on big data, machine learning, and applications*. Springer.

Meyer, D. N. (2016). How exactly does word 2 vec work? Presented at the.

Mikolov, T., Chen, K., Corrado, G., & Dean, J. (2013). Efficient estimation of word representations in vector space. arXiv preprint arXiv:1301.3781.

Moran, S., et al. (2016). Enhancing first story detection using word embeddings. In *Proceedings of the 39th international ACM SIGIR conference on Research and development in information retrieval*. https://doi.org/10.1145/2911451.2914719

Panagiotou, N., Akkaya, C., Tsioutsiouliklis, K., Kalogeraki, V., & Gunopulos, D. (2016). First story detection using entities and relations. In *COLING 2016, the 26th International Conference on Computational Linguistics: Technical Papers* (pp. 3237–3244).

Petrović, S., Osborne, M., & Lavrenko, V. (2010). Streaming first story detection with application to Twitter. In *Human language technologies: The 2010 annual conference of the north american chapter of the association for computational linguistics* (pp. 181–189).

Petrovic, S., Osborne, M., & Lavrenko, V. (2010). Streaming first story detection with application to Twitter. In *Proc. of NAACL*.

Plank, B., Søgaard, A., & Goldberg, Y. (2016). Multilingual part-of-speech tagging with bidirectional long short-term memory models and auxiliary loss. In *54th Annual Meeting of the Association for Computational Linguistics ACL 2016 - Short Paper* (pp. 412–418). https://doi.org/10.18653/v1/p16-2067

Ramos, J. (2003). Using TF-IDF to determine word relevance in document queries. In *Proceedings of the first instructional conference on machine learning*.

Repp, Ø. K., & Ramampiaro, H. (2016). Event detection in social media - detecting news events from the Twitter stream in real-time, *93*.

Sabbah, T., et al. (2017). Modified frequency-based term weighting schemes for text classification. *Applied Soft Computing, 58*, 193–206. https://doi.org/10.1016/j.asoc.2017.04.069

Soliman, A. B., Eissa, K., & El-Beltagy, S. R. (2017). AraVec: A set of Arabic word embedding models for use in Arabic NLP. *Procedia Computer Science, 117*, 256–265. https://doi.org/10.1016/j.procs.2017.10.117

Twitter formal website. https://about.twitter.com/company

Vinhkhuc: JFastText. (2016). https://github.com/vinhkhuc/JFastText. Accessed 12 Dec 2016.

Vogiatzis, M. (2020). How to spot first stories on Twitter using Storm. https://micvog.com/2013/09/08/storm-first-story-detection/#more-125. Accessed 15 Jan 2020.

Vosoughi, S., Vijayaraghavan, P., & Roy, D. (2016). Tweet2vec: Learning tweet embeddings using character-level cnn-lstm encoder-decoder. In *Proceedings of the 39th international ACM SIGIR conference on Research and development in information retrieval* (pp. 1041–1044). ACM.

Vuurens, J. B. P., & de Vries, A. P. (2016). First story detection using multiple nearest Neighbors. In *Proceedings of the 39th international ACM SIGIR conference on Research and development in information retrieval* (pp. 845–848). ACM.

Wieting, J., Bansal, M., Gimpel, K., & Livescu, K. (2016). Charagram: Embedding words and sentences via character n-grams. In *EMNLP 2016 - Conf. Empir. Methods Nat. Lang. Process. Proc.* (pp. 1504–1515). https://doi.org/10.18653/v1/d16-1157

Yang, Y., Pierce, T., & Carbonell, J. (1998). A study of retrospective and on-line event detection. In *Proceedings of the 21st annual international ACM SIGIR conference on research and development in information retrieval* (pp. 28–36). ACM.

Yu, X., & Vu, N. T. (2017). Character composition model with convolutional neural networks for dependency parsing on morphologically rich languages. In *ACL 2017 - Proceedings of the 55th Annual Meeting of the Association for Computational Linguistics (Long Paper 2)* (pp. 672–678). https://doi.org/10.18653/v1/P17-2106

Chapter 10
Using a Bi-Directional Long Short-Term Memory Model with Attention Mechanism Trained on MIDI Data for Generating Unique Music

Ashish Ranjan, Varun Nagesh Jolly Behera, and Motahar Reza

10.1 Introduction

"Beauty is in the eye of the beholder," is a common phrase used in English. This phrase is especially appropriate when we consider something like art (Kumar and Garg, 2010). Art is something very human by nature, and artistic judgements are highly subjective, entirely based on the individual observer and his/her experiences (Johnston and Franklin, 1993). For example, it is hard for some people to accept rap as a form of music, and as a form of art (Shusterman, 1991). Art always has a pattern that is recognizable, while still containing some irregularities (Field and Golubitsky, 2009). When that mixture of patterns and irregularities is able to please us, we start to like that form of art.

Human beings have long believed that art represents a creative activity where machines may never rival humans (Varshney et al., 2013). However, the advent of powerful computers and state-of-the-art machine learning algorithms raises the possibility of a prominent role for AI art in the future (Besold et al., 2015). Since machines do not need rest, and an AI could generate thousands of art pieces per hour, there would be increased efficiency and cost-effectiveness in art-based sectors (Toivonen, 2020). Generative adversarial networks (GAN) (Goodfellow et al., 2014) are very popular these days, especially for creating hyper-realistic faces (Gauthier, 2014). These systems have huge face datasets to learn from Gao et al. (2007), Kasinski et al. (2008), and Ma et al. (2015). An example of such a system is

A. Ranjan · V. N. Jolly Behera (✉)
R&D Division, Just Another Media Laboratory (JAM Lab), Mumbai, India
e-mail: ashish.ranjan@jamlab.in; varun.behera@jamlab.in

M. Reza
Department of Mathematics, GITAM University Hyderabad Campus, Hyderabad, India
e-mail: mreza@gitam.edu

M. Alloghani et al. (eds.), *Artificial Intelligence for Data Science in Theory and Practice*, Studies in Computational Intelligence 1006,
https://doi.org/10.1007/978-3-030-92245-0_10

DeepFake (Korshunova et al., 2017), which superimposes the generated face over a subject's face in an image or video so that it seems like some other person is present in the image or video. Another example is neural style transfer (Li et al., 2017), which learns the art style from one image and applies the style to another image, effectively transferring the style. This technique has been used to recreate the art style of famous painters (Gatys et al., 2016).

A more technical term for machine-produced art is computational creativity (CC) (Colton and Wiggins, 2012). CC is an interdisciplinary field spanning topics like AI, cognition, art, design, and philosophy (Minsky, 1967). This field of study focuses on creating algorithms that mimic human creativity, understand human creativity, mathematically define creativity, or create systems that assist and enhance the human creative process (Gero, 2000). Music is probably one of the toughest computational creativity problems, as it is difficult to analyze with a computer (Leach and Fitch, 1995). It is easier to analyze an image, as there is no temporal component, and the whole image can be observed at once. Whereas in the case of music, the temporal dimension, and the fact that it is highly rhythmic yet coherent, poses greater challenges for the computer. Generating melodic audio that sounds realistic is difficult and is an open problem in the field of machine learning (Papadopoulos and Wiggins, 1999). Imagine a DJ who not only mixes music but also composes it spontaneously. It could also potentially be used to generate music based on the current mood of the user, or even replicate the style of famous musicians to generate new music. The possibilities are endless.

This chapter explores a methodology for generating unique music that is melodic using deep learning. The methodology involves a bi-directional long short-term memory recurrent neural network with an attention mechanism. This system learns the melodic and rhythmic patterns and dependencies in order to generate novel music as output. MIDI (Moog, 1986), a popular high-level representation for music, was used.

10.2 Algorithmic Composition

Algorithmic composition refers to the process of using some parameters and combining parts of music to form a whole (Nierhaus, 2009). Generally algorithmic composition is either music composed by a machine or music created with the help of a machine. However, it is not necessarily computer-based, an example being Mozart's musical dice game (Hedges, 1978).The idea of algorithmic composition has in fact been around for centuries, as far back as Pythagoras suggesting music and mathematics are the same (Fauvel et al., 2006). Some techniques include species counterpoint, d'Arezzo's Micrologus (Sullivan, 1989), Schoenberg's twelve-tone technique (Dahlhaus, 1987), or Cage's aleatoric music (Jeongwon and Hoo, 2002).

One of the first computational models used for generating music utilized Markov chains (McAlpine et al., 1999). This and other algorithms such as fractals (Hsü and Hsü, 1991), statistical models (Conklin, 2003), L-systems (Worth and Stepney,

2005) can be applied to random real-world data to generate music. These algorithms all take advantage of the mathematical patterns inherent in music. Algorithmic composition is not restricted to computational means, as represented by above-mentioned methods. Different categories of algorithmic music may be described as follows:

1. **Mathematical models**: These refer to stochastic processes, and more specifically Markov chains, used for real-time algorithmic composition, due to their low complexity. An example of such a model used for algorithmic composition is Cybernetic Composer (Ames and Domino, 1992).
2. **Knowledge based systems**: These are systems that use symbols and specified constraints to achieve algorithmic composition. An example of this is CHORAL (Ebcioğlu, 1988). Building systems like these for music is difficult and time-consuming.
3. **Grammar**: A grammar set for a musical language can be defined, and its rules used randomly, to generate unique music. An augmented transition network (ATN) (Woods, 1970) is used to accomplish this task. The music generated may not be of optimal quality, as the semantics defined may not be robust.
4. **Evolutionary methods**: Genetic algorithms (GA) (Whitley, 1994) are used in this approach. They use either an objective fitness function or a human fitness function. Since music is subjective, the latter may produce undesirable results (Horner and Goldberg, 1991).
5. **Systems which learn**: These consist of machine learning-based systems, especially artificial neural networks (Browne, 2001). This paper explores the usage of a long short-term memory based learning system (Mangal et al., 2019).
6. **Translational models**: Information from non-musical media is translated into musical media. For example, using time series data generally used for load forecasting, and instead restructuring the dimensions as pitch and scale.
7. **Hybrid systems**: Using any of these systems in combination forms a hybrid system. Basically an ensemble model for algorithmic composition.

The focus of this chapter is on current AI techniques, specifically techniques related to natural language processing (NLP).

10.3 Related Work

Automatic music generation may be considered as a subfield of automated audio processing, which is a burgeoning field that has researchers all around the world producing new ideas. WaveNet (van den Oord et al., 2016) is a prominent example of a successful innovation in this area. WaveNet is a probabilistic autoregressive generative model that can generate raw audio. It was developed by DeepMind Technologies (www.deepmind.com) and has been used in text-to-speech applications. The output produced by WaveNet is very realistic and creates voices that are almost human-like. Its effectiveness can be summarized by its high mean opinion score

(MOS) given when it was first introduced, and the fact that it has been implemented as the voice of Google Assistant. It is not just limited to text-to-speech applications. When the developers trained it with a musical dataset, realistic sounding rhythmic audio, which can easily be identified as music, was generated. The parallel version of WaveNet is very fast (van den Oord et al., 2018). It can generate 20 seconds of audio in just 1 second. Since it is a convolutional neural network, training it is very easy for the computer, as opposed to a recurrent neural network, which is generally suited to handle sequences of data such as audio.

The authors of (Blaauw and Bonada, 2017) created a "neural parametric singing synthesizer" (NPSS) that uses a modified version of WaveNet to generate music based on features produced by a parametric vocoder. NPSS is capable of producing very high-quality results, as expected from WaveNet.

Recurrent neural networks (RNNs) have also been used for the purpose of music generation. One such application can be seen in the paper "LSTM Based Music Generation System" (Mangal et al., 2019). Since audio samples at any instant are highly dependent on previous samples for coherent audio, RNNs, especially long short-term memory (LSTM), can be used to take care of these dependencies. This approach works well with classical and jazz music because those are more free-form and without definite song structure like verse, chorus, and bridge. It only needs to know the recent context of notes to generate the next note.

Generative adversarial networks (GANs) can also be used for the purpose of music generation. One such system is "WGANSing" (Chandna et al., 2019), a multi-voice singing voice synthesizer based on the Wasserstein-GAN. It employs two networks. One is called the generator, which generates new music and tries to fool the other network, and the other is called the discriminator, which tries to discriminate between real musical recording and generated music. After each epoch of training, both networks get better at their jobs and the generator can be used to generate music.

The artificial intelligence company OpenAI (openai.com) has also worked in this field and created MuseNet. It uses the same general-purpose unsupervised technology as GPT-2 (Radford et al., 2019), which is a large-scale transformer model trained to predict the next token in a sequence, whether audio or text. It is able to extend music whose starting fragment is given as input, or it can generate music from scratch. OpenAI's blog (openai.com/blog/musenet/) gives an interactive demonstration of MuseNet. The demo enables the user to select a starting song fragment, target style, and instrumentation from which MuseNet generates an extended musical passage.

Some companies such as Jukedeck (purchased in 2019 by TikTok), Aiva (aiva.ai), and Ampermusic (ampermusic.com) have commercialized their music generation systems, which their customers use to generate custom royalty-free music.

Frameworks such as Magenta (magenta.tensorflow.org) are freely available for researchers to work in the field of generating art by computer. Magenta, in particular, is an open-source research project exploring the role of machine learning as a tool in the creative process. It is available for Python and JavaScript and was built using TensorFlow. Magenta includes utilities for manipulating source data (primarily music and images), using this data to train machine learning models, and finally generating new content from these models.

Reference Dieleman et al. (2018) explores autoregressive discrete autoencoders (ADAs) as a means to enable autoregressive models such as WaveNet to capture long-range correlations in waveforms, which otherwise captures local signal structures. The use of autoencoders in this aspect has also been demonstrated by the "Neural Composer" which is available from the YouTube channel "Code Parade." Neural Composer uses principal component analysis to identify the top 40 principal components and assigns them to control sliders to generate custom music.

10.4 The Problem of Music Generation

Music is an art form with a strong emotional component. Researchers have identified 13 key emotional dimensions evoked by music: amusement, joy, eroticism, beauty, relaxation, sadness, dreaminess, triumph, anxiety, scariness, annoyance, defiance, and feeling pumped up (Cowen et al., 2020). This research aims to generate realistic music in the sense that it can evoke a similar range of feelings and attain similar levels of satisfaction. In order to achieve this goal, the problem statement may be divided into several subtasks.

1. The system should be able to generate new audio using high-level representation such as MIDI.
2. The system should learn the rhythmic sequences of music during the training phase and produce similar but novel music as output. For audio, each sample depends on the context of the previous samples. This is even more true in the case of music. Generally, the context of the music samples depends on the context of all of the previous samples. This is best implemented in WaveNet and made computationally efficient by using dilated convolutions, which makes large skips in the data to get a larger receptive view of the data.
3. Based on a set of input parameters (e.g., the principal component amplitudes in the Neural Composer), the system should be able to produce appropriate results. When these parameters are changed in various combinations, original new music with different characteristics is generated.
4. Overall, the output should be melodic. (this is the main goal of the research.)
5. The output should have a decent mean opinion score (MOS). Since the evaluation of music is highly subjective, the best way to check the performance of a music generation system is to conduct a survey in which human participants try to guess if music being played has been generated by a computer or if it is an actual human performance.

Some optional requirements are:

1. Parameters such as mood, genre, instrument, mode, key, tempo, etc. could be set as inputs. This poses difficulties because it requires extensive labeled datasets.
2. A future objective could be to integrate a lyrics generator and a vocal synthesizer.

10.5 Dataset

Some freely available digital music datasets that have been identified are:

1. **The MAESTRO dataset**: MIDI and audio edited for synchronous tracks and organization (MAESTRO) Hawthorne et al. (2018) is a dataset composed of over 200 hours of paired audio (in WAV format) and MIDI recordings from ten years of the International Piano-e-Competition. Audio and MIDI files are aligned within 3 ms accuracy and sliced to individual musical pieces, which are annotated with composer, title, and year of performance. Uncompressed audio is of CD quality or higher (44.1–48 kHz 16-bit PCM stereo). The WAV files comprise 103 GB (122 GB uncompressed), while the MIDI files comprise 57 MB (85MB uncompressed).
2. **The Million Song dataset**: The core of this dataset (Bertin-Mahieux et al., 2011) is the feature analysis and metadata for one million songs, provided by the company The Echo Nest. The dataset does not include any audio, only the derived features. Note, however, that sample audio can be fetched from services like 7digital (7digital.com) using code. Songs are given tags such as artist, release date, album, etc. Labels such as danceability, key, mode, tempo, etc. are also given, and these labels are usable for our purposes. The size of the dataset is 280 GB (the full dataset) or 1.8 GB (1%, or 10,000 songs).
3. **MagnaTagATune dataset**: This dataset (Law and Von Ahn, 2009) contains the most versatile collection of labels. Unfortunately the labels are often unorganized and duplicated, such as a song being labeled as both "classic" and "classical" genres. Extensive preprocessing will be required to work with this dataset. It can be used for genre, instrument, and mood classification.

If an adequate dataset for the user's objectives do not exist, a custom dataset may be created. However, this requires time and effort to manually label the data. Generally, the format of digital music datasets is either WAV, which is directly in the audio domain, or MIDI, which directly encodes note times, pitches, and durations which may be interpreted by a synthesizer. MIDI is generally easier to use for music generation purposes.

10.6 Data Representation

In order to generate music, it is necessary to understand how it is represented digitally. Music is generally represented in the following ways:

10.6.1 Notes

A note is the symbolic representation of a musical tone (Strunk, 1942). Figure 10.1 shows the conventional state notation for an ascending series of notes. Notes are represented in English by the letters A through G. Another popular representation is Do, Re, Mi, Fa, So, La, and Ti, where Do = C, Re = D, … Ti = B. Notes may also be raised or lowered a semitone using the symbols # and ♭, respectively. The octave (eighth note) is the same as the first with double the frequency value, meaning it has a higher pitch. In addition to the note symbol, written notes may also have a note value. This note value determines the duration of the notes.

The frequency of these notes is measured in hertz (Hz), as mechanical systems produce them. There are 12 notes with fixed frequencies, defined around the central note A_4. This note has a frequency of 440 Hz.

10.6.2 Raw Audio

Uncompressed audio stored in raw form is called raw audio (Tzanetakis and Cook, 2000). Figure 10.2 shows the waveform. Raw audio files have no header and hence cannot be played without some user input. Thus file formats like WAV and AIFF are generally used, as they are lossless and are nearly the same size. Other lossless formats include FLAC, ALAC, and WMA Lossless. The most popular of these are the WAV and FLAC formats.

Waveform Audio File Format (WAV) is an audio file format based on the Resource Interchange File Format (RIFF) bit-stream format. It is similar to Audio Interchange File Format (AIFF), as both are based on RIFF. It is typically uncompressed and uses linear pulse-code modulation for bit-stream encoding. Although it is possible to store compressed audio using the WAV format, the WAV format is

Fig. 10.1 Data representation: musical notes

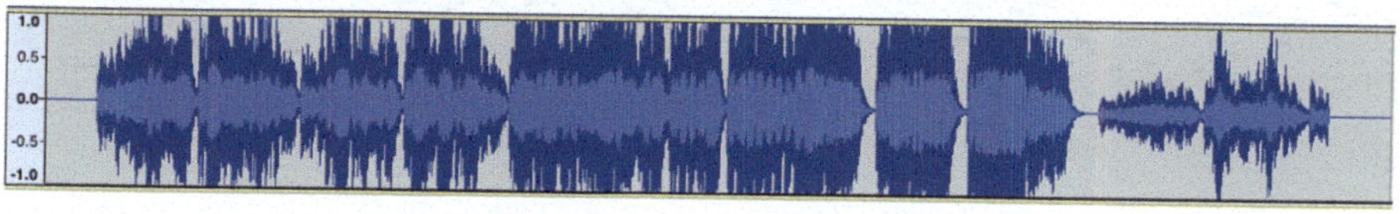

Fig. 10.2 Data representation: raw audio

generally used when the best audio quality is required. Typically audio is encoded at 44.1 kHz and a 16-bit sample rate.

Free Lossless Audio Codec (FLAC) is a lossless format for storing audio. This codec allows for lossless compression ranging from 50 to 70%. It can be easily decoded and streamed.

10.6.3 MIDI

Musical Instrument Digital Interface (MIDI) (Moog, 1986) is a universal protocol/interface for many electronic instruments and computers. It allows for ease in recording various instruments and simultaneously interfacing them via computer. The file format for this interface is called Standard MIDI File (SMF) and usually has a .mid extension. These files are very small and can be easily transferred between various devices. Actual audio is not stored in SMF files, which only store information such as pitch, duration, and loudness of notes, as well as global parameters such as tempo and instrumentation. This is the reason for the files' extremely small sizes (usually under 1 megabyte) and high portability. Music in MIDI format can also be easily tweaked and/or entirely transformed: for example, instrumentation, pitch, and tempo can all be adjusted in seconds. As a result, MIDI is generally used for constructing soundtracks. Figure 10.3 shows the graphical representation of MIDI.

10.6.4 Piano Rolls

The original "piano rolls" (Shi et al., 2017) were rolls of papers with holes that represent data for notes, as can be seen in Fig. 10.4. These piano rolls provided inspiration for MIDI and are the logical predecessors. In software, "piano rolls" generally refer to the visual representation of MIDI data. It is theoretically possible to use this visual data as input for some image-based generation models, like GANs. This may be explored in future work.

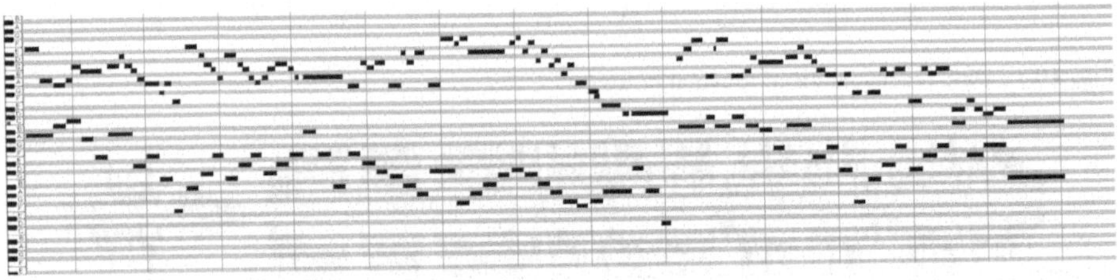

Fig. 10.3 Data representation: MIDI

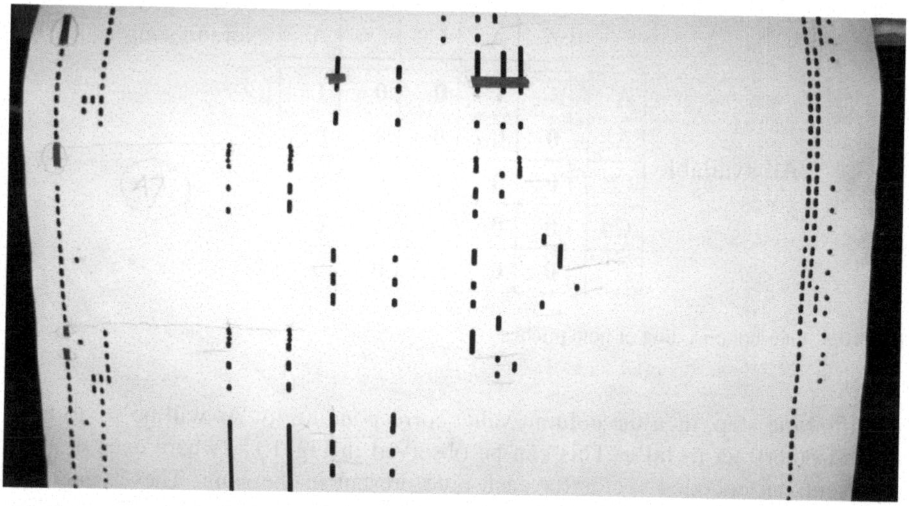

Fig. 10.4 Data representation: traditional piano roll, which resembles the presentation of MIDI data

10.7 Preparing the Data

For our implementation, a custom data set comprising of 92 MIDI files was used, consisting of songs from modern video game soundtracks. The MIDI files were interpreted and instrumentalized using the Python package Music21.

10.7.1 Dataset Description

The dataset contains 92 songs. All of the songs are from video game soundtracks and have an average length of 3–4 minutes. Standard MIDI encoding digitizes note volumes to integer values from 0 to 127.

10.7.2 Data Encoding

As MIDI files cannot be directly used in an LSTM, they need to be encoded into an appropriate input vector. For this purpose, a one-hot encoding methodology is used. For each note in the MIDI file, an array of Boolean values with a length equal to the number of available notes is created. Each element of the array corresponds to the available notes. For a note, all values of the array are initialized to false (or 0), except for the corresponding note which is initialized to true (or 1). This array is taken as an input vector to the model. For example, if the note "A" is present at a

All available notes / Note in song	A	A	C	C#	A
A	1	1	0	0	1
A#	0	0	0	0	0
C	0	0	1	0	0
C#	0	0	0	1	0
E	0	0	0	0	0

Fig. 10.5 One-hot encoding of note pitches

specific time step, then the column value corresponding to "A" will be set to true, and all others set to false. This can be observed in Fig. 10.5, where each column represents an encoded vector for each note present in the song. These generated vectors become the input to the proposed model. Each column in Fig. 10.5 is input to each bi-directional LSTM unit (explained in the next section).

10.7.3 Data Loading

The MIDI files are stored in a directory on the hard disk and are read in Python using the Music21 package. Each MIDI file is converted into one sequence of notes to be input to the model. These sequences are stored in a list which form the entire dataset that will be given as input to the model.

10.8 Proposed Model

Since music is highly rhythmic and dependent on previous notes, it can be thought of as a sequence of notes played at certain intervals. The intricate interdependence of notes suggests the use of a neural network model. To handle time dependencies a RNN is needed, and LSTM is used to handle long-term dependencies, as described below.

10.8.1 Long Short-Term Memory

Traditional neural networks assume each input is independent of each other. But in the current scenario, each input is dependent on the previous inputs through time. This can be solved using RNNs. However, a regular RNN has a serious drawback

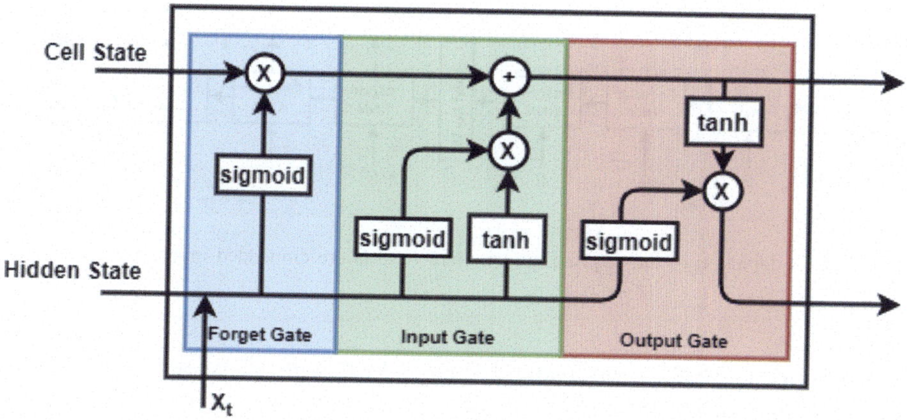

Fig. 10.6 An LSTM cell, where x_t is the input at time t, $\times$ is multiplication and $+$ is addition

in that as the length of the input sequence increases, the contribution of the earlier inputs in predicting subsequent input drops significantly. This happens so drastically that the inputs that were given long before have a negligible effect on the prediction. This is the vanishing gradient problem, or the problem with long-term dependencies.

This is solved using a special type of RNN, called long short-term memory (LSTM). An LSTM consists of units which are in turn composed of cells, where each cell consists of a memory (which holds the cell state) and three types of gates: forget gates, input gates, and output gates. The forget gate combines hidden information (output from previous cells) with new incoming information and produces an output that regulates the level of "forgetting" of the previous cell state. This is done by passing the combined hidden+new information through a sigmoid function which produces a value between 0 and 1, where 0 indicates "forget everything" and 1 indicates "remember everything." Next, the input gate updates the cell state utilizing hidden and new information, as well as the current cell state. The update is accomplished by rescaling new plus hidden information to the interval $[-1, 1]$ using the tanh function, multiplying by a forgetting factor, and adding the result to the forget gate's output times the current cell state. Finally, the output gate is used compute the cell's output by multiplying the rescaled current state times another forgetting factor. With this structure, the LSTM cell is able to store long-term information in its cell state, while suppressing currently irrelevant information in its output. An LSTM cell is shown in Fig. 10.6.

10.8.2 Sequence-to-Sequence Generation Model

Since music is a sequence of notes, the model needs to learn from a sequence of notes and generate a sequence of notes. This can be achieved using the sequence-to-

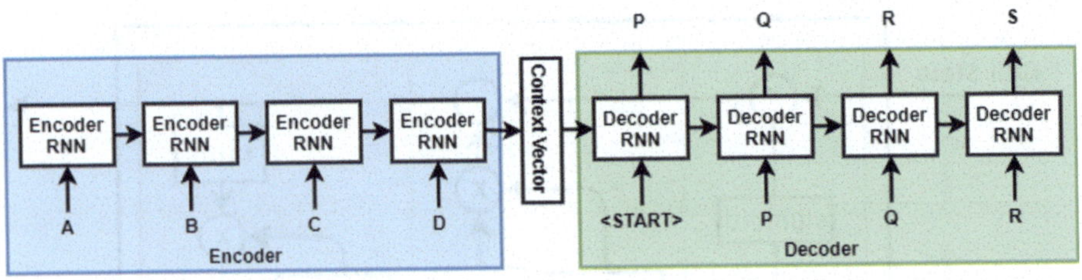

Fig. 10.7 A depiction of the seq2seq model. Here A to D are embedded inputs and P to S are outputs

sequence model seq2seq, originally developed by Google (Sutskever et al., 2014). It has two parts: an encoder and a decoder. The encoder uses a series of LSTM units to learn from a sequence of inputs to get a context vector, which is the product of the learning process. Each LSTM unit is provided with one element of the sequence of inputs. The context vector is then used by the decoder to generate the output sequence one note at a time by predicting a starting output and taking the output of the previous LSTM unit as input for the next. Seq2seq is able to generate one sequence from another without running into the vanishing gradient problem. This is a great advantage. An overall depiction of seq2seq is shown in Fig. 10.7.

10.8.3 Attention Mechanism

In the real world, in order to make inferences it is necessary to focus on what is important. For example, if a description of an image is required, we focus on the activity being shown in the image or the participants involved and ignore the unimportant details. For example, an image of an apple contains not only just the apple but also contains where the apple is placed, what the lighting conditions are, etc. However, if we are interested in the condition of the apple we do not pay attention to these other details and focus on the apple. The attention mechanism (first introduced by Bahdanau et al., 2014) allows an encoder-decoder to focus on what is important in the current scenario. It calculates alignment scores corresponding to each input to determine which inputs are more important than others.

10.8.4 Combining Seq2seq with Attention Mechanism

A seq2seq model with attention mechanism (Vaswani et al., 2017) can be used for music generation. This is because seq2seq is able to generate one sequence from another without running into the vanishing gradient problem. It uses an encoder layer and a decoder layer. The encoder layer converts an input into its corresponding hidden vector and context using bi-directional LSTM units, while the decoder reverses these operations. Using just seq2seq is not enough, as it gives equal importance to every note in the sequence of notes. In realistic music, more emphasis is given to the chorus and other sections of the music as they may be repeated multiple times. To solve this issue, an attention layer is added between the encoder and decoder layers. The attention layer helps remember short as well as long-term dependencies and helps determine which notes must be given more importance.

The Bahdanau attention mechanism is used with the seq2seq model. The encoder layer consists of a bi-directional LSTM layer and the decoder layer uses a regular LSTM layer. The encoded input vector is converted to hidden states by the encoder. This is fed into the attention layer, which produces a context vector. Finally, the context vector thus produced will be used to decode the output sequence. The model is shown in Fig. 10.8. Its specifications can be seen in Fig. 10.9. Under "Layer (type)," "bidirectional_1" corresponds to the encoder layer, "seq_self_attention_1" corresponds to the attention layer, and "cu_dnnlstm_2" corresponds to the decoder layer.

The training is done in order for the model to learn the patterns in music created by humans. For music generation after training, a random note or a sequence of

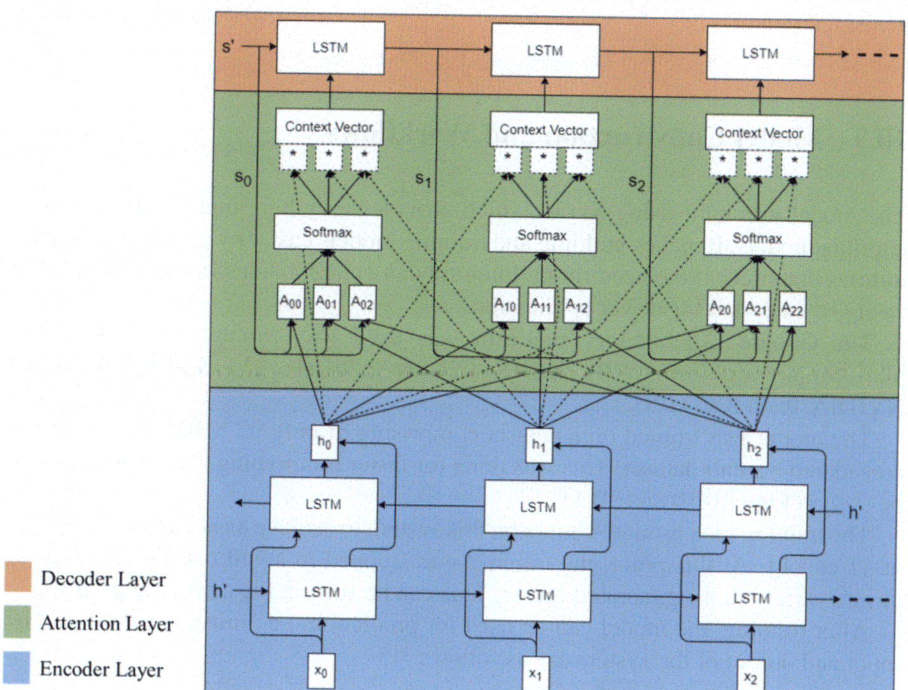

Fig. 10.8 Proposed model. Here x_t is the input at time t, h_t is the hidden state at time t, and s is the cell state of decoder LSTM units

```
Layer (type)                 Output Shape              Param #
=================================================================
bidirectional_1 (Bidirection (None, 100, 1024)         2109440
_________________________________________________________________
seq_self_attention_1 (SeqSel (None, 100, 1024)         65601
_________________________________________________________________
dropout_1 (Dropout)          (None, 100, 1024)         0
_________________________________________________________________
cu_dnnlstm_2 (CuDNNLSTM)     (None, 100, 512)          3149824
_________________________________________________________________
dropout_2 (Dropout)          (None, 100, 512)          0
_________________________________________________________________
flatten_1 (Flatten)          (None, 51200)             0
_________________________________________________________________
dense_1 (Dense)              (None, 2826)              144694026
_________________________________________________________________
activation_1 (Activation)    (None, 2826)              0
=================================================================
Total params: 150,018,891
Trainable params: 150,018,891
Non-trainable params: 0
```

Fig. 10.9 Keras model specifications

notes is given as input to the model to predict the next note. This is repeated by the model to generate the next notes and finally, the produced sequence is output as the generated music.

10.9 Model Construction and Workflows

The model was built using Keras, which provides a Python interface for machine learning models. It makes building and training models easier by providing modular utilities that prevent the need for building fundamental modules from scratch. It can easily be extended for newer applications.

The Google Colaboratory platform was used for training the model. This platform provides an Intel(R) Xeon(R) CPU @ 2.20GHz, 12 GB of RAM, and an NVIDIA Tesla K80 GPU.

The model was trained on a dataset comprising of only 92 MIDI files, which is considered a small dataset. This was done for faster prototyping. The workflow can be seen in Figs. 10.10 and 10.11.

The training took around 4 hours on this system to achieve an accuracy of 93.02% at 31 epochs. At this point, the training was stopped to avoid overfitting, so as to ensure variety in the generated music. This can be seen in Figs. 10.12 and 10.13.

After training, the model can be used for prediction, i.e., music generation. The input and output of the system are as follows.

1. **Input**: A random note or a sequence of notes.
2. **Output**: The input extended to a length of nearly two minutes. This sequence is rhythmic.

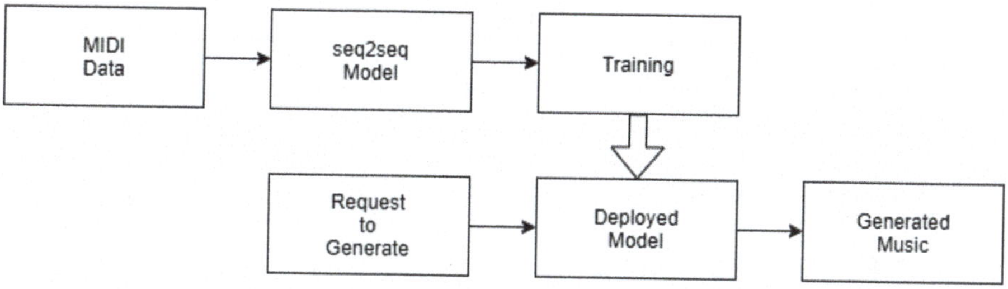

Fig. 10.10 ML workflow

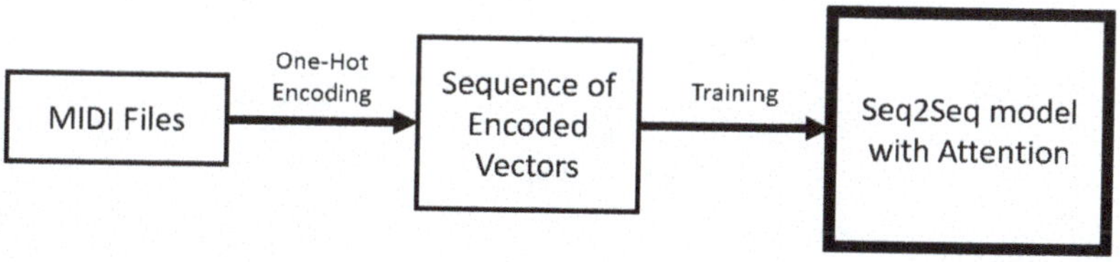

Fig. 10.11 Training workflow

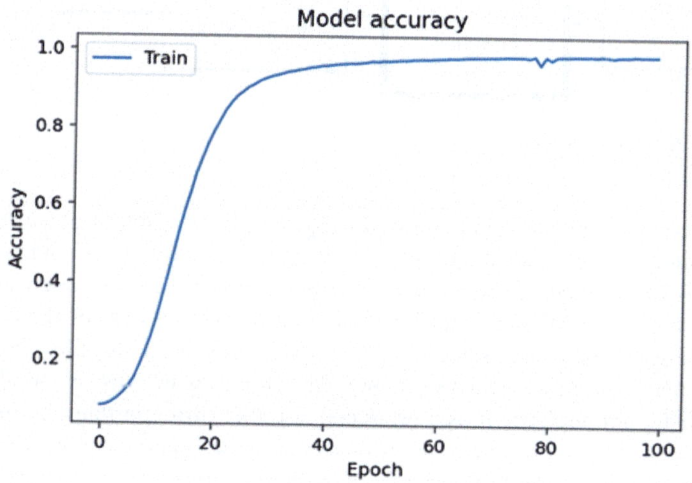

Fig. 10.12 Model accuracy

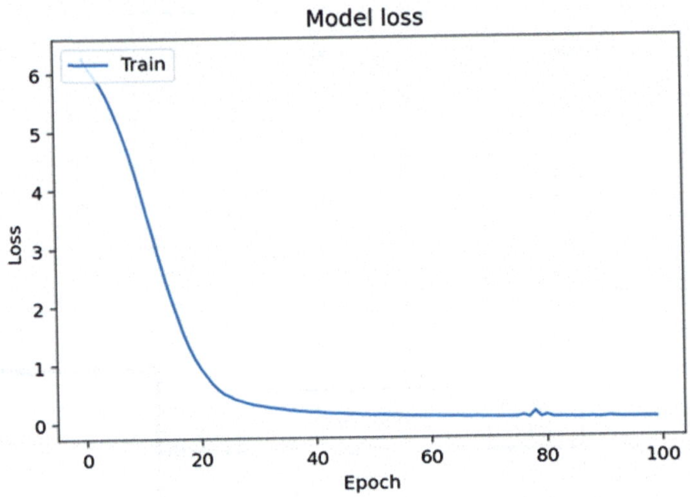

Fig. 10.13 Model loss

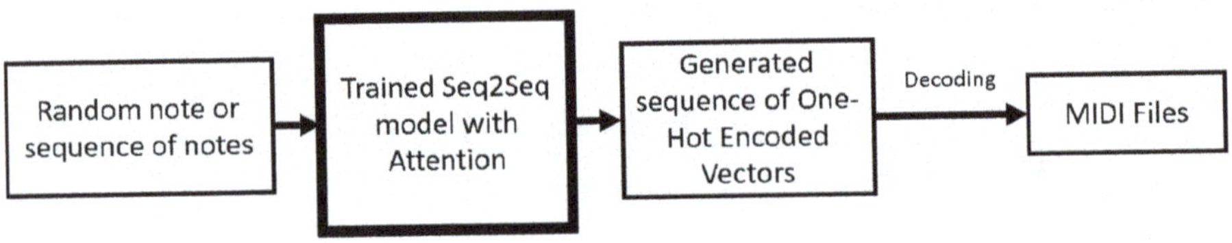

Fig. 10.14 Prediction workflow

The prediction workflow is shown in Fig. 10.14. The output generation takes approximately 4–5 seconds to complete. This means it is highly efficient.

The model produces a sequence of one-hot encoded vectors which need to be converted back into MIDI format. To do this, a reverse process of the initial one-hot encoding can be done. Since the produced results are audio files, no image is produced and cannot be shown on paper. The model generates results that sound fairly realistic although the music produced is rather different than the input. Still, some nuances of the input songs can be observed in the generated music. This means we can tell that the model has learnt from the input songs.

The results can be improved by training on a much larger dataset. In our attempt, we trained on only 92 songs for faster prototyping, but future research can be extended to huge datasets as well.

In generating new music based on a determined dataset, this valid question may arise: What if the same starting note is used multiple times? The answer is it will most likely result in the same output, but this can be solved easily. The input can instead be extended to a sequence of notes. A sequence of notes allows for space for many random combinations of starting notes.

10.10 Model Evaluation

The performance of this model was evaluated using a mean opinion score (MOS) test. 100 subjects were chosen for the survey, and 5 random songs, generated by the model, were played for each of them. The subjects were asked to rate the songs on a five-point scale (1:Very Poor, 2:Poor, 3:Average, 4:Good, 5:Very Good). A total of 500 different songs were evaluated this way, and a mean of the scores was evaluated. Figure 10.15 summarizes the results of the survey. "Good" was the most common ranking, although "Average" also received many votes, while "Very Good" and "Poor" received nearly equal numbers of votes. The overall average MOS was 3.4, while the standard deviation was 1.1. Despite the substantial fraction of average and below-average ratings, nonetheless the results are a promising indication of the potential of generated music. One possibility is to use generated compositions as inspiration for new music rather than being used directly.

For better availability, the entire system described in this chapter has been converted into a web app, accessible at https://research.jamlab.in/music-generator. The web app has a highly intuitive interface as can be seen in Fig. 10.16. When the user presses the "Generate Song" button, the model will be invoked and a MIDI file will be generated. The generated MIDI file can be listened to using the MIDI player on the page itself. While the music file is playing, the notes are represented with a virtual piano which (like an old-fashioned piano roll) depresses the keys corresponding to notes as they are played.

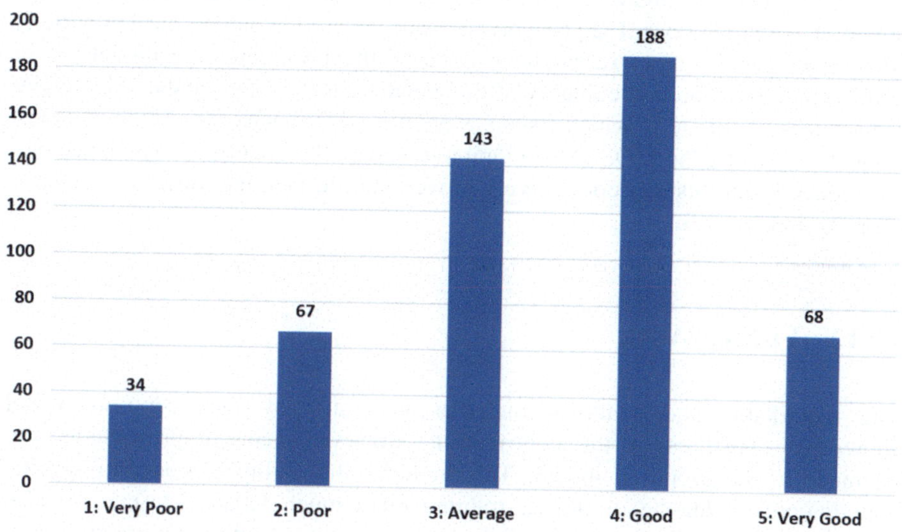

Fig. 10.15 MOS or mean opinion score (y-axis shows the number of song rankings)

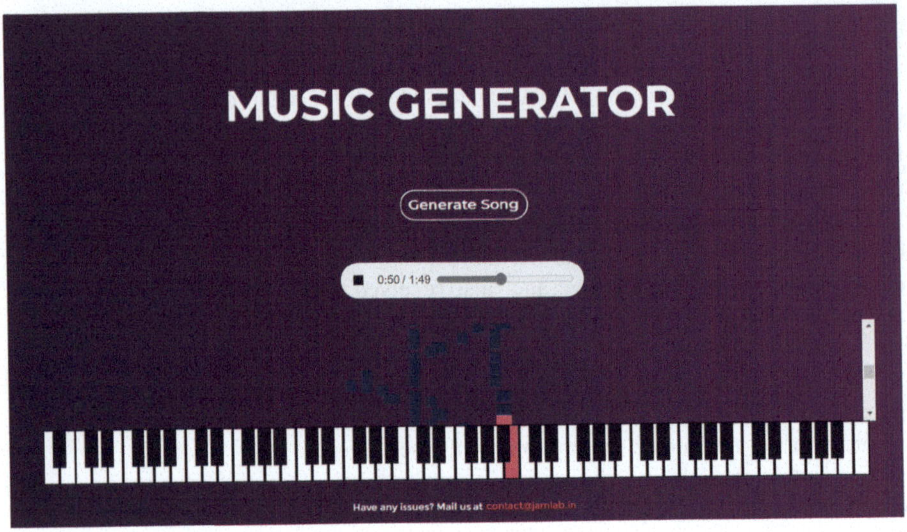

Fig. 10.16 Web application

10.11 Conclusion

A deep learning model capable of generating a specific type of music by learning from MIDI data was created. The model uses a bi-directional LSTM model with attention. Although several seq2seq based music generation models have previously been developed by other researchers, none of them validates the model for its quality. In this research, the quality of the music produced was validated by means of an extensive survey, in which 5 songs were played for each of the 100 human test subjects. By using the MOS (Mean Opinion Score), the quality of experience was calculated. From this it is conclusively proven that, in fact, the music generated is pleasing and beautiful.

10.12 Future Work

The remarkable and beautiful melodies generated by the proposed model show the robustness of the contribution. However, many improvements can be made to the proposed model. This includes the ability to generate specific genre/mood/style/theme of music, training on a much larger dataset, and also exploring the possibility of having different instruments and multiple tracks.

The piano roll based approach is also something that can be explored. Finally, a future model that has a text generator able to generate lyrics of a specific genre/mood/style/theme, and a voice synthesizer able to sing the lyrics based on the tune generated, shall be explored. This will lead to a fully independent music production system capable of disrupting the music industry and bringing a new era of completely free and unlicensed music.

Acknowledgments We thank Prof. Rahul Roy (GITAM Hyderabad) and Prof. Debananda Kanhar (NIST, Berhampur) for their useful discussions and feedback. We thank our supervisor Dr. Motahar Reza (GITAM Hyderabad) for his continued encouragement and guidance. We thank Snehal Swadhin (IIT Kharagpur) for building the frontend and deploying the web app. We thank Sachidananda Padhiali and Shubham Kardam (JAM Lab) for their support.

References

Ames, C., & Domino, M. (1992). Cybernetic Composer: An overview. In *Understanding Music with AI: Perspectives on Music Cognition. 1992*, 186–205.

Bahdanau, D., Cho, K., & Bengio, Y. (2014). Neural machine translation by jointly learning to align and translate. arXiv:1409.0473.

Blaauw, M., & Bonada, J. (2017). A neural parametric singing synthesizer modeling timbre and expression from natural songs. *Applied Sciences, 7*(12), 1313.

Browne, C. B. (2001). System and method for automatic music generation using a neural network architecture. U.S. Patent No. 6,297,439. 2 Oct. 2001.

Bertin-Mahieux, T., Ellis, D. P. W., Whitman, B., & Lamere, P. (2011). The Million Song Dataset. In *Proceedings of the 12th International Society for Music Information Retrieval Conference (ISMIR 2011)* (pp. 591–596).

Besold, T. R., Schorlemmer, M., & Smaill, A. (Eds.). (2015). *Computational creativity research: Towards creative machines*. Amsterdam: Atlantis Press.

Chandna, P., Blaauw, M., Bonada, J., & Gomez, E. (2019). WGANSing: A multi-voice singing voice synthesizer based on the Wasserstein-GAN. In *2019 27th European Signal Processing Conference (EUSIPCO)*. Piscataway: IEEE.

Colton, S., & Wiggins, G. A. (2012). Computational creativity: The final frontier? *Frontiers in Artificial Intelligence and Applications, 242*, 21–26.

Conklin, D. (2003). Music generation from statistical models. In *Proceedings of the AISB 2003 Symposium on Artificial Intelligence and Creativity in the Arts and Sciences.*

Cowen, A. S., Fang, X., Sauter, D., & Keltner, D. (2020). What music makes us feel: At least 13 dimensions organize subjective experiences associated with music across different cultures. *Proceedings of the National Academy of Sciences, 117*(4), 1924–1934.

Dahlhaus, C. (1987). *Schoenberg and the new music: Essays by Carl Dahlhaus*. Cambridge: Cambridge University Press.

Dieleman, S., van den Oord, A., & Simonyan, K. (2018). The challenge of realistic music generation: Modelling raw audio at scale. In *Advances in Neural Information Processing Systems*.

Ebcioğlu, K. (1988). An expert system for harmonizing four-part chorales. *Computer Music Journal, 12*(3), 43–51.

Fauvel, J., Flood, R., & Wilson, R. J. (Eds.) (2006). *Music and mathematics: From Pythagoras to fractals*. Oxford: Oxford University Press on Demand.

Field, M., & Golubitsky, M. (2009). *Symmetry in chaos: A search for pattern in mathematics, art, and nature*. Philadelphia: Society for Industrial and Applied Mathematics.

Gao, W., Cao, B., Shan, S., Chen, X., Zhou, D., Zhang, X., & Zhao, D. (2007). The CAS-PEAL large-scale Chinese face database and baseline evaluations. *IEEE Transactions on Systems, Man, and Cybernetics-Part A: Systems and Humans, 38*(1), 149–161.

Gatys, L. A., Bethge, M., Hertzmann, A., & Shechtman, E. (2016). Preserving color in neural artistic style transfer. arXiv:1606.05897

Gauthier, J. (2014). Conditional generative adversarial nets for convolutional face generation. Class Project for Stanford CS231N: Convolutional Neural Networks for Visual Recognition, Winter semester 2014.5, 2.

Gero, J. S. (2000). Computational models of innovative and creative design processes. *Technological Forecasting and Social Change, 64*(2–3), 183–196.

Goodfellow, I., Pouget-Abadie, J., Mirza, M., Xu, B., Warde-Farley, D., Ozair, S., Courville, A., & Bengio, Y. (2014). Generative adversarial nets. In *Advances in Neural Information Processing Systems*.

Hawthorne, C., Stasyuk, A., Roberts, A., Simon, I., Huang, C.-Z. A., Dieleman, S., Elsen, E., Engel, J., & Eck, D. (2018). Enabling factorized piano music modeling and generation with the MAESTRO dataset. arXiv:1810.12247.

Hedges, S. A. (1978). Dice music in the eighteenth century. *Music & Letters, 59*(2), 180–187.

Horner, A., & Goldberg, D. E. (1991). *Genetic algorithms and computer-assisted music composition*. (vol. 51). Ann Arbor: Michigan Publishing, University of Michigan Library.

Hsü, K. J., & Hsü, A. (1991). Self-similarity of the "1/f noise" called music. *Proceedings of the National Academy of Sciences, 88*(8), 3507–3509.

Jeongwon, J., & Hoo, S. S. (2002). Roland Barthes Text'and aleatoric music: Is 'the birth of the reader' the birth of the listener? *Muzikologija, 2*, 263–281.

Johnston, V. S., & Franklin, M. (1993). Is beauty in the eye of the beholder? *Ethology and Sociobiology, 14*(3), 183–199.

Kasinski, A., Florek, A., & Schmidt, A. (2008). The PUT face database. *Image Processing and Communications, 13*(3–4), 59–64.

Korshunova, I., Shi, W., Dambre, J., & Theis, L. (2017). Fast face-swap using convolutional neural networks. In *Proceedings of the IEEE International Conference on Computer Vision*.

Kumar, M., & Garg, N. (2010). Aesthetic principles and cognitive emotion appraisals: How much of the beauty lies in the eye of the beholder? *Journal of Consumer Psychology, 20*(4), 485–494.

Law, E., & Von Ahn, L. (2009). Input-agreement: A new mechanism for collecting data using human computation games. In *Proceedings of the SIGCHI Conference on Human Factors in Computing Systems*.

Lawrence, S., Giles, C. L., Tsoi, A. C. & Back, A. D. (1997). Face recognition: A convolutional neural-network approach. *IEEE Transactions on Neural Networks, 8*(1), 98–113.

Leach, J., & Fitch, J. (1995). Nature, music, and algorithmic composition. *Computer Music Journal, 19*(2), 23–33.

Li, Y., Wang, N., Liu, J., and Hou, X. (2017). Demystifying neural style transfer. arXiv:1701.01036.

Ma, D. S., Correll, J., & Wittenbrink, B. (2015). The Chicago face database: A free stimulus set of faces and norming data. *Behavior Research Methods, 47*(4), 1122–1135.

Mangal, S., Modak, R., & Joshi, P. (2019). LSTM based music generation system. arXiv:1908.01080.

McAlpine, K., Miranda, E., & Hoggar, S. (1999). Making music with algorithms: A case-study system. *Computer Music Journal, 23*(2), 19–30.

Minsky, M. (1967). Why programming is a good medium for expressing poorly understood and sloppily formulated ideas. In *Design and planning II-computers in design and communication* (pp. 120–125).

Moog, R. A. (1986). Midi: Musical instrument digital interface. *Journal of the Audio Engineering Society, 34*(5), 394–404.

Nierhaus, G. (2009). *Algorithmic composition: Paradigms of automated music generation*. Berlin: Springer.

Papadopoulos, G., & Wiggins, G. (1999). AI methods for algorithmic composition: A survey, a critical view and future prospects. In *AISB Symposium on Musical Creativity, Edinburgh* (vol. 124).

Radford, A., Wu, J., Child, R., Luan, D., Amodei, D., & Sutskever, I. (2019). Language models are unsupervised multitask learners. *OpenAI Blog, 1*(8), 9.

Shi, Z., Arul, K., & Smith, III, J. O. (2017). Modeling and digitizing reproducing piano rolls. In *18th International Society for Music Information Retrieval Conference, Suzhou, China, ISMIR'17*.

Shusterman, R. (1991). The fine art of rap. *New Literary History, 22*(3), 613–632.

Strunk, O. (1942). The tonal system of Byzantine music. *The Musical Quarterly, 28*(2), 190–204.

Sullivan, B. (1989). Interpretive models of guido of Arezzo's micrologus. *Comitatus: A Journal of Medieval and Renaissance Studies, 20*(1), 20–42.

Sutskever, I., Vinyals, O., & Quoc V. Le. (2014). Sequence to sequence learning with neural networks. In *Advances in Neural Information Processing Systems*.

Toivonen, H. (2020). Computational creativity beyond machine learning. *Physics of Life Reviews, 34–35*, 52–53.

Tzanetakis, G., & Cook, P. (2000). Audio information retrieval (AIR) tools. In *Proceedings of International Symposium on Music Information Retrieval*.

van den Oord, A., Dieleman, S., Zen, H., Simonyan, K., Vinyals, O., Graves, Al., Kalchbrenner, N., Senior, A., & Kavukcuoglu, K. (2016). Wavenet: A generative model for raw audio. arXiv:1609.03499.

van den Oord, A., Li, Y., Babuschkin, I., Simonyan, K., Vinyals, O., Kavukcuoglu, K., van den Driessche, G., Lockhart, E., Cobo, L. C., Stimberg, F., Casagrande, N., Grewe, D., Noury, S., Dieleman, S., Elsen, E., Kalchbrenner, N., Zen, H., Graves, A., King, H., Walters, T., Belov, D., & Hassabis, D. (2018). Parallel wavenet: Fast high-fidelity speech synthesis. In *International Conference on Machine Learning. PMLR*.

Varshney, L. R., Pinel, F., Varshney, K. R., Schörgendorfer, A., & Chee, Y.-M. (2013). Cognition as a part of computational creativity. In *2013 IEEE 12th International Conference on Cognitive Informatics and Cognitive Computing*. Piscataway: IEEE.

Vaswani, A., Shazeer, N., Parmar, N., Uszkoreit, J., Jones, L., Gomez, A. N., Kaiser, L., & Polosukhin, I. (2017). Attention is all you need. In *Advances in Neural Information Processing Systems*.

Whitley, D. (1994). A genetic algorithm tutorial. *Statistics and Computing, 4*(2), 65–85.

Woods, W. A. (1970). Transition network grammars for natural language analysis. *Communications of the ACM, 13*(10), 591–606.

Worth, P., & Stepney, S. (2005). Growing music: Musical interpretations of L-systems. In *Workshops on Applications of Evolutionary Computation*. Berlin: Springer.

Index

M. Alloghani et al. (eds.), *Artificial Intelligence for Data Science in Theory and Practice*, Studies in Computational Intelligence 1006,
https://doi.org/10.1007/978-3-030-92245-0

The manufacturer's authorised representative in the EU is Springer Nature Customer Service Centre GmbH, Europaplatz 3, 69115 Heidelberg, Germany. If you have any concerns regarding our products, please contact ProductSafety@springernature.com

Printed and bound by CPI Group (UK) Ltd, Croydon, CR0 4YY
22/09/2026
02220961-0006